Natural Selection

MONOGRAPHS IN POPULATION BIOLOGY

EDITED BY ROBERT M. MAY

1. The Theory of Island Biogeography, by Robert H. MacArthur and Edward O. Wilson
2. Evolution in Changing Environments: Some Theoretical Explorations, by Richard Levins
3. Adaptive Geometry of Trees, by Henry S. Horn
4. Theoretical Aspects of Population Genetics, by Motoo Kimura and Tomoko Ohta
5. Populations in a Seasonal Environment, by Stephen D. Fretwell
6. Stability and Complexity in Model Ecosystems, by Robert M. May
7. Competition and the Structure of Bird Communities, by Martin
8. Sex and Evolution, by George C. Williams
   L. Cody
9. Group Selection in Predator-Prey Communities, by Michael E. Gilpin
10. Geographic Variation, Speciation, and Clines, by John A. Endler
11. Food Webs and Niche Space, by Joel E. Cohen
12. Caste and Ecology in the Social Insects, by George F. Oster and Edward O. Wilson
13. The Dynamics of Arthropod Predator-Prey Systems, by Michael P. Hassell
14. Some Adaptations of Marsh-Nesting Blackbirds, by Gordon H. Orians
15. Evolutionary Biology of Parasites, by Peter W. Price
16. Cultural Transmission and Evolution: A Quantitative Approach, by L. L. Cavalli-Sforza and M. W. Feldman
17. Resource Competition and Community Structure, by David Tilman
18. The Theory of Sex Allocation, by Eric L. Charnov
19. Mate Choice in Plants: Tactics, Mechanisms, and Consequences, by Mary F. Willson and Nancy Burley
20. The Florida Scrub Jay, by G. E. Woolfenden and J. W. Fitzpatrick
21. Natural Selection in the Wild, by John A. Endler

Natural Selection in the Wild

JOHN A. ENDLER

PRINCETON UNIVERSITY PRESS

PRINCETON, NEW JERSEY

1986

Published by Princeton University Press, 41 William Street,
Princeton, New Jersey 08540
In the United Kingdom: Princeton University Press,
Chichester, West Sussex

Library of Congress Cataloging-in-Publication Data

Endler, John A., 1947–
Natural selection in the wild.

(Monographs in population biology: 21
Bibliography: p.
Includes indexes.
1. Natural selection. I. Title. II. Series.
QH375.E53 1985 575.01´62 85-42683

ISBN 0-691-08386-X ISBN 0-691-08387-8 (pbk.)

This book has composed in Lasercomp Baskerville

Princeton University Press books are printed on acid-free
paper and meet the guidelines for permanence and
durability of the Committee on Production Guidelines for
Book Longevity of the Council on Library Resources

Printed in the United States of America by Princeton Academic Press

9 8 7 6

TO

Arthur J. Cain

AND

Bryan C. Clarke

THIS BOOK IS AFFECTIONATELY
DEDICATED

Contents

Preface

This is a book on natural selection for those who are interested in both the evolution and ecology of natural populations of organisms. Natural selection is an immense and important subject, yet there have been few attempts to summarize its effects on natural populations, and fewer still that are concerned with the methods and problems of working with natural selection in the wild. This book attemps to permeate this void. Theoreticians will find nothing that is new in this book, and those who primarily work with laboratory populations will look in vain for their favorite examples. Laboratory populations serve as models of nature, and help to test specific predictions or conjectures about the way nature works; but without extensive knowledge of natural selection in the wild, we have no idea how relevant experiments or theory are to the evolution of natural populations. This is meant to be a field guide rather than a textbook or manual, so common methods will not be given in detail (major references are always provided).

Although this book is not meant to be a eulogy to Eris, I expect that nobody will find it wholly satisfactory. Among the many people who have read it in manuscript, some find parts exceedingly helpful, while others find the very same parts boring or superfluous. To many, the most irritating parts will probably be found in Chapters 1, 2, and 8 because they attempt to put the various points-of-view, definitions, and meanings of natural selection in perspective, and everyone thinks that his own emphasis is most important. A typical reaction is: "I find it fascinating that more than 100 years after the *Origin* and Mendel there can be two major positions on this everyday phrase"; actually there are more than two (Figure 1.2).

A major problem in this subject is that there is a multiplicity of meanings for the same terms, and the same terms mean different things to different people. This complexity and our understanding of natural selection may benefit from my crude attempt at standardization. I hold no hope of convincing those who are used to a particular meaning of a word to conform to the suggested standardization. For example, in another book (Endler 1977), I clearly distinguished migration, dispersal, and gene flow—terms that had a similar mix-up with, say, fitnes and adaptation (Chapter 2). Migration can have nothing to do with gene flow, and has a meaning entirely different to ecologists and ethologists compared to population geneticists, yet the latter insist on calling gene flow "migration." This is perfectly clear if one is sufficiently narrow to read only the literature within a minutely circumscribed field. However, it is a source of serious confusion to those with more catholic tastes and to beginning students. That the need for standardization is urgent is shown by the latest example of synonomy in name only: selection gradients. In 1973 I used the term "selection gradient" to denote a geographic gradient in natural selection, and this followed general usage going back to Huxley and others in the 1930s (references in Endler 1977). More recently Lande and others (for example, Lande and Arnold 1983) used "selection gradient" in a completely different sense: the rate of change of fitness with trait value. This comes from a mathematical model of natural and artificial selection of multiple traits, which can be regarded as a surface with a gradient in value. This is a perfectly correct usage of the term "selection gradient" and would be fine if the word had not already been used for geographically varying selection for the past fifty years! This is a potentially serious problem for the unwary because a cline may be associated with both kinds of "selection gradients." We need some sort of rule of priority, analogous to that in systematics, which states that the earlier meaning of the word should stand, and another word should be used for the new meaning. However, human nature being what it is, this rule will probably be ignored. I therefore present the standardization attempt (Chapters 1 and 2) primarily as a guide for students and others who read the literature outside a narrow field of specialization. The main purpose of this book is not to argue about words, but to summarize natural selection as it can be seen in natural populations.

I thank Ric Charnov for encouraging me to write, and to finish, this book. Elliott Sober very kindly let me read his splendid book (1984) before publication. I am grateful for discussions and comments on all or parts of the manuscript from Steve Arnold, Robert Brandon, Arthur Cain, Ric Charnov, Bryan Clarke, Blaine Cole, Jim Crow, Monica Geber, Michael Ghiselin, Peter Grant, Paul Harvey, Lynn Jorde, Mark Kirkpatrick, Russ Lande, Yan Linhart, Jim Mallett, Bob May, Tracy McLellan, Trevor Price, Sam Skinner, Elliott Sober, Bill Stubblefield, Sam Sweet, and David Temme. They certainly have not always agreed with my observations and conclusions.

December 1984

Natural Selection in the Wild

CHAPTER ONE

Introduction

> Of the great principles of truth which the first speculatists discovered, the simplicity is embarrassed by ambitious additions, or the evidence obscured by inaccurate argumentation; and as they descend from one succession of writers to another, like light transmitted from room to room, they lose their strength and splendour, and fade at last in total evanescence. The systems of learning therefore must be sometimes reviewed, complications analyzed into principles, and knowledge disentangled from opinion.
>
> Samuel Johnson, *The Rambler*, 14 September 1751

Natural selection is a major part of the theory of evolution (Darwin 1859; Fisher 1930; Mayr 1963; Ghiselin 1969), yet there is much argument and confusion as to what it is, what it is not, and even whether or not it exists.[1] These disputations have tended to befog the larger questions of mechanisms and even the validity of the theory of evolution (Ghiselin 1969; Wassermann 1981a,b; Gould 1982). It is the purpose of this book to describe natural selection clearly, show that it is neither a tautology nor a metaphysical exercise, discuss the problems of its demonstration and measurement, present the critical evidence for its existence, and place it in perspective. This chapter will define natural selection, relate it to genetic drift and evolution, discuss the restricted meanings the term "natural selection" often takes, and summarize some of its modes.

[1] For differing views and arguments, see Cox 1981; Flew 1981; Gendron 1981; Pearson 1981; Robson 1981; Stephenson 1981; Wasserman 1981a,b.

Throughout this book, lengthy lists of references are given in footnotes, while shorter lists remain parenthetically cited in the text. This dual citation system is not intended to draw any distinction between references, but only to make the text easier to read.

1.1. DEFINITION OF NATURAL SELECTION

Natural selection can be defined as a *process* in which:

If a population has:

a. variation among individuals in some attribute or trait: *variation*;

b. a consistent relationship between that trait and mating ability, fertilizing ability, fertility, fecundity, and, or, survivorship: *fitness differences*;

c. a consistent relationship, for that trait, between parents and their offspring, which is at least partially independent of common environmental effects:[2] *inheritance*.

Then:

1. the trait frequency distribution will differ among age classes or life-history stages, beyond that expected from ontogeny;

2. if the population is not at equilibrium, then the trait distribution of all offspring in the population will be predictably different from that of all parents, beyond that expected from conditions *a* and *c* alone.

Conditions *a*, *b*, and *c* are necessary and sufficient for the process of natural selection to occur, and these lead to deductions *1* and *2*. As a result of this process, but not necessarily, the trait distribution may change in a predictable way over many generations.[3] The process of natural selection has been called a law (Reed 1981) because if the initial conditions are fulfilled, the conclusions necessarily follow; the principle behind the law is a syllogism. Natural selection probably should not be called a biological law. It proceeds not for biological reasons, but from

[2] The environment common to parents and offspring can yield a correlation between parents and offspring if there is an environmental component to trait variation, the environment is heterogeneous, and there is a physical association between parents and offspring (Falconer 1981).

[3] Modified after Fisher 1930, Falconer 1981, Bulmer 1980, Ewens 1979, Ghiselin 1981, and Williams 1970, 1973.

the laws of probability; conditions *a–c* contain the only biological content.

1.2. RELATIONSHIP TO GENETIC DRIFT AND EVOLUTION

Genetic drift is a random sampling process of alleles between generations. The necessary and sufficient conditions for genetic drift (Wright 1931, 1942; Kimura 1983; Lande 1976a, 1980) differ in only two respects from those for natural selection (Table 1.1): (1) condition *b* is absent (by definition), and (2) the effective population size must be small enough to ensure that sampling error is significant. Of course it is perfectly possible for both natural selection and genetic drift to occur simultaneously in small populations. We can divide both processes into phenotypic difference (conditions *a* and *b*) and genetic response (*c*); then the only difference is that the phenotypic difference is consistent (in sign) among generations during natural selection, but randomly varying during genetic drift (Table 1.1). The distinction is blurred where selection varies at random in time.

Evolution may be defined as any net directional change or any cumulative change in the characteristics of organisms or populations over many generations—in other words, descent with modification (after Lincoln et al. 1982). It explicitly includes the *origin* as well as the *spread* of alleles, variants, trait values, or character states. Evolution may occur as a result of natural selection, genetic drift, or both (Figure 1.1); the minimum requirements are those for either process (Table 1.1). Natural selection does not necessarily give rise to evolution, and the same is true for genetic drift.

By definition, a population at equilibrium has the same trait distribution at each generation; it is not evolving. This may result solely from natural selection, or through a combination of natural selection and other countervailing evolutionary factors. If a population is not at equilibrium, then evolution can proceed, and this was the main interest of Darwin and the other

TABLE 1.1. The relationships among natural selection, genetic drift, and evolution

Property	Natural Selection	Genetic Drift	Evolution[a]
	Necessary and Sufficient Conditions		
Condition *a* (trait variation)	required	required	required
Condition *b* (fitness differences)	required	absent (by definition)	not required (more likely if present)
Condition *c* (inheritance)	required	required	required
Small effective population size	not required	required	not required unless condition *b* is absent
Origin of new variation	not required	not required	required
	Other Properties		
Observed differences among phenotypes or age classes (condition *b*)	consistent in time (deduction 1)	random in time	consistent (or relatively so) in time
Stable equilibrium possible?	yes (part of deduction 2)	yes[b]	no[c] (by definition)
Measure of degree of condition *b*	fitness[d]	(observed "fitness")	durability[d]

[a] In the sense of any cumulative change in the characteristics of a population over many generations (see text).
[b] Larger populations may appear to be at equilibrium. Also, a stable equilibrium or stationary distribution is possible with mutation and/or gene flow.
[c] A stable equilibrium is possible, but once it is reached, evolution has stopped (by definition) until conditions change.
[d] See Chapter 2 for discussion.

early evolutionists. But if a population is at equilibrium, no evolution is possible (by definition) unless the relationships in conditions *a–c* change, or some other evolutionary factors come into play. Whether or not a population is at equilibrium when studied depends upon its history as well as on current conditions *a–c*.

Population geneticists use a different definition of evolution: a change in allele frequencies among generations. This meaning is quite different from the original; it now includes random as well as directional changes (more than the shaded part of Figure 1.1), but it does not require the origin of new forms. It is roughly equivalent to microevolution (subspecific evolution; macroevolution involves major trends, or transspecific evolution; see Rensch 1959, Mayr 1963). Unfortunately, the use of the population genetics definition often results in an overemphasis on changes in allele frequencies and an underemphasis on (or no consideration of) the *origin* of the different alleles and their properties. Both are important in evolution (see Chapter 8). An

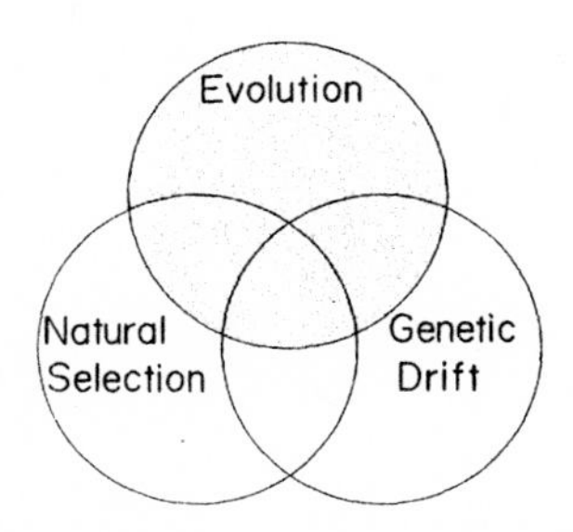

FIGURE 1.1. A Venn diagram, showing the relationship between the necessary and sufficient conditions for natural selection, genetic drift, and evolution. The shaded area results in cumulative changes in genotype frequencies or trait values, and the shaded area outside the domains of natural selection and genetic drift may be due to mutation, meiotic drive, and other processes. See also Table 1.1. The relative sizes and degree of overlap are not meant to imply anything about the relative importance or frequency of these phenomena. Evolution can be regarded either as the shaded area or as the area of all three circles (see text); the former is used in this book.

additional problem is that, for quantitative genetic traits, the frequencies of alleles at many contributing loci can change while the overall mean and variance of the trait remain roughly constant. In this book I will use the older definition of evolution (Lincoln et al. 1982; Figure 1.1, shaded), rather than the population-genetics definition. For either definition, natural selection is sufficient for evolution, but it is neither necessary for nor does it guarantee evolution.

The origin of conditions *a*, *b*, and *c* is an issue separate from natural selection. Natural selection takes these conditions as given, and it results in consequences 1 and 2. These consequences may or may not affect the conditions for natural selection in the next generation. The conditions are a joint effect of the environment, the genetic system, and the history of the population, and may evolve as a result of many different factors. Thus the origin of conditions *a*, *b*, and *c* is a function of genetics, evolution and ecology, not necessarily of natural selection. We will return to this in Chapters 2 and 8. Natural selection must not be equated with evolution, though the two are intimately related.

1.3. RESTRICTED MEANINGS OF "NATURAL SELECTION"

The term "natural selection" means different things to different people, and this often leads to confusion in the literature. Three restricted meanings are relatively common, and they partially overlap: mortality selection, nonsexual selection, and phenotypic selection. The last two are parts of distinctions which are of very great theoretical importance, and have significantly increased our understanding of natural selection and evolution.

The restricted meanings can be placed in a broader perspective. By the nature of its definition, natural selection can be broken down into various components in two different and independent ways, depending upon alternate subprocesses differing in dynamics and outcome (Figure 1.2A), or component or sequential subprocesses (Figure 1.2B). The restricted meanings of "natural selection" are parts of these subdivisions: mortality and nonsexual selection are in the first subdivision, while phenotypic selection is in the second.

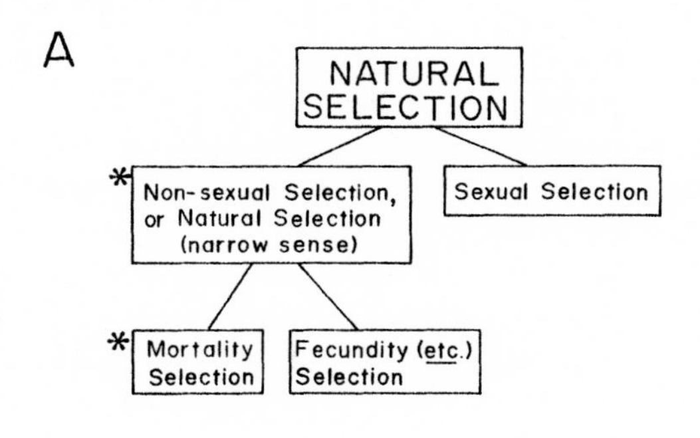

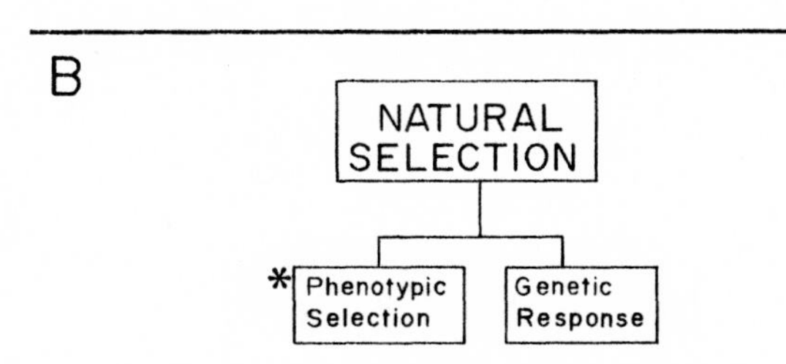

FIGURE 1.2. Two logical and independent subdivisions of natural selection, illustrating the restricted meanings that "natural selection" often takes in the literature. *A*, subdivision by alternate subprocesses; *B*, subdivision by sequential subprocesses. The subprocesses in *A* are distinguished with respect to condition *b* (fitness differences) and implicitly contain conditions *a* (trait variance) and *c* (inheritance) for natural selection; each is necessary and sufficient for natural selection to occur. Each does not guarantee evolution because conditions *a*, *b*, and *c* can result in a stable equilibrium. The subprocesses in *B* are separated on the basis of the three conditions for natural selection: phenotypic selection requires conditions *a* and *b*, while the genetic response requires condition *c*. Neither is sufficient by itself for the process of natural selection, though both together are necessary and sufficient. Asterisks indicate the subprocesses that are frequently called "natural selection" in the literature.

1.3.1. Mortality Selection

Natural selection is sometimes used to refer only to the effects of consistent phenotype-specific mortality (discussed in Fisher 1930 and Ghiselin 1969; see Figure 1.2A); examples are Hailman (1982) and Darlington (1983). Consider an expanding population consisting of two genotypes, one of which is increasing faster than the other. Some researchers do not consider this a case of natural selection because there is no mortality (Hailman 1982; pers. comm. 1983). Restriction to mortality selection in the literature depends, in part, on equating natural selection with "survival of the fittest." It also depends upon a singular interpretation of what Darwin meant by natural selection. Three quotes from Darwin (1859) give slightly different impressions:

> Owing to this struggle for life, any variation, however slight and from whatever cause proceeding, if it be in any degree profitable to an individual of any species, in its infinitely complex relations to other organic beings and to external nature, will tend to the preservation of that individual, and will generally be inherited by its offspring. The offspring, also, will thus have a better chance of surviving, for, of the many individuals of any species which are periodically born, but a small number can survive. I have called this principle, by which each slight variation, if useful, is preserved, by the term of Natural Selection, in order to mark its relation to man's power of selection. (p. 61)

> I should premise that I use the term Struggle for Existence in a large and metaphorical sense, including dependence of one being on another, and including (which is more important) not only the life of the individual, but success in leaving progeny. (p. 62)

> If such do occur, can we doubt (remembering that many more individuals are born than can possibly survive) that

> individuals having any advantage, however slight, over others, would have the best chance of surviving and procreating their kind? On the other hand, we may feel sure that any variation in the least degree injurious would be rigidly destroyed. This preservation of favourable variations and the rejection of injurious variations, I call Natural Selection. (pp. 80–81)

These definitions might be understood to indicate only mortality effects, but they are easier to interpret as including other components such as fecundity and fertility. Further reading of Darwin (1859 and 1871) suggests that he thought natural selection included more than mortality. Mortality selection is clearly a special case of natural selection (Figure 1.2A); it is too restricted to be useful except in special cases, although natural selection will proceed with mortality selection alone.

1.3.2. Sexual and Nonsexual Selection

Darwin (1859, p. 88; 1871) made a careful distinction between natural selection and sexual selection (Figure 1.2A): sexual selection is a result of differential mating success, including fertilization and pairing. The distinction was made because traits favored by sexual selection may sometimes be disadvantageous, or opposed by other components of natural selection (Darwin 1871; Ghiselin 1974; Wade and Arnold 1980). Thus the outcome, as well as the dynamics, can be quite different from what Darwin and many biologists would regard as "natural selection" (Fisher 1930; Lande 1981; Kirkpatrick 1982; Arnold 1983a). Explicit as well as implicit differences of opinion abound on whether or not sexual selection is a subset of natural selection; in addition, to add to the confusion, other aspects of differential reproductive success (such as fertility) have often also been included with sexual selection. Because the consequences of mating success are so distinct, it is best to restrict "sexual selection" to its original meaning and its application to differential mating success, rather than to include all aspects of reproductive success (Ghiselin 1974; Wade and Arnold 1980; Arnold 1983a).

As defined in this book, sexual selection is a logical subset of the process of natural selection. This is true because (1) mating ability is one of several alternatives in condition *b*, and (2) the definition of the process takes no account of the details of its outcome; it merely states that the trait frequency distribution may change if conditions *a*, *b*, and *c* are met. In fact there is no difference between sexual and nonsexual selection in the methods of demonstration or measurement (see Chapter 6 and Arnold and Wade 1984a,b). In addition, other components of natural selection can oppose one another. In the very general sense, sexual selection is a subset or aspect of natural selection, but at a lower level (considering their dynamics and outcomes) they are very distinct. Perhaps we should use the term "organic selection" for the general process, sexual selection for processes involving mating success, and natural selection (narrow sense) for the remainder. However, for simplicity in this book the term "natural selection" will be used for the general process, and nonsexual selection and sexual selection will be used for the specific subprocesses. Note that mortality selection is a special case of nonsexual selection (Figure 1.2A), and, like mortality selection, sexual and nonsexual selection are sufficient by themselves for natural selection to proceed.

1.2.3. Phenotypic Selection and Response

This subdivision of natural selection is independent of the previous ones (Figure 1.2B). Quantitative geneticists and animal breeders decompose the process of natural selection into phenotypic selection and genetic (or "evolutionary") response (Fisher 1930; Haldane 1954; Falconer 1981; Lande and Arnold 1983). Phenotypic selection is the *within-generation* change in the trait distribution among cohorts (or the difference between the actual number of mates and the effective number of mates in the case of sexual selection), and is independent of any genetic system or genetic determination. In terms of the definition of natural selection, phenotypic selection requires conditions *a* and *b*. The response is the genetic change that occurs as a result of phenotypic selection in combination with the genetic system, which requires condition *c*. This is a very important and useful distinction (see Falconer 1981).

If there is no inheritance (condition *c*) the process of natural selection cannot occur. In spite of this, phenotypic selection is sometimes called "natural selection" (for example, Lande and Arnold 1983). One good reason for this is that natural selection works on phenotypes and not on genotypes (Mayr 1963; Lewontin 1974). But natural selection is the differential survival and perpetuation of phenotypes, and perpetuation requires inheritance. Phenotypic selection determines the distribution of traits during reproduction, but inheritance is required to transform the distribution into the next generation. To say that natural selection is synonymous with phenotypic selection is to trivialize it—this is tantamount to saying that there are differences among different phenotypes, which can easily lead to tautology (Chapter 2).

The restriction of natural selection to phenotypic selection results at least in part from an inconsistent distinction between evolution, natural selection, and genetic drift. It also accounts for the occasional use of the term "evolutionary response" for "genetic response." "Evolutionary response" is an unfortunate usage because natural selection does not necessarily result in evolution—at equilibrium there can be a genetic response to phenotypic selection every generation, but no change in trait distributions, that is, no evolution (Table 1.1). Random genetic drift can also yield differences among age classes, which will appear to be phenotypic selection if only a few generations are examined. This apparent phenotypic selection will be followed by a genetic response as the random within-generation change is transformed into the next generation through the hereditary process, but that is not natural selection! It is condition *b* and not *c* that distinguishes natural selection from genetic drift (Table 1.1); merely splitting off *c* is insufficient. To be logically consistent, we must either include genetic response as part of the process of natural selection (as in the definition in this book), or distinguish *three* processes: (1) phenotypic ("natural") selection; (2) genetic response; and (3) cumulative genetic change (evolution). Phenotypic selection and genetic response should be regarded as subprocesses of natural selection. This also appears to be closer to what Darwin intended:

> But if variations useful to any organic being do occur, assuredly individuals thus characterised will have the best chance of being preserved in the struggle for life; and from the strong principle of inheritance they will tend to produce offspring similarly characterised. This principle of preservation, I have called, for the sake of brevity, Natural Selection. (Darwin 1859, p. 127)

Natural selection can preserve differences; this is impossible without condition *c*. An additional reason for including *c* as a requirement for natural selection is that its separation from *a* and *b* has been a major contributing factor in keeping the fields of ecology and genetics separate.

To put this usage into a broader perspective, those who restrict "natural selection" to phenotypic selection also call natural selection, as defined in this book, "evolution"; those who are more careful call it "evolution by natural selection." But evolution is more than merely a change in trait distributions or allele frequencies; it also includes the *origin* of the variation. I will return to this in Chapter 8. For these reasons, the distinctions between natural selection, genetic drift, and evolution as shown in Table 1.1 will be used in this book.

As with the distinction between sexual and nonsexual selection, the distinction between phenotypic selection and genetic response is an important and useful one, and is based upon a subdivision of the process of natural selection (Figure 1.2B), but unlike the first subdivision (Figure 1.2A), phenotypic selection (conditions *a* and *b*) and genetic response (condition *c*) are not by themselves sufficient for the *process* of natural selection.

To reiterate, natural selection may be broken down into two orthogonal subdivisions (Figure 1.2), and the restricted meanings emphasize these subdivisions. The first subdivision (Figure 1.2A) breaks the process into complete components, differing only in the details of condition *b*; natural selection can occur for any one of the subprocesses in this subdivision. On the other hand, the second subdivision (Figure 1.2B) breaks the process sequentially into one containing conditions *a* and *b* and one with *c*; natural selection cannot occur without both subprocesses of the second subdivision. Mortality and nonsexual selection are components of the first subdivision, while phenotypic selection is a component of the second; of the three restricted meanings, only phenotypic selection is insufficient by itself for natural selection to proceed. If mortality, nonsexual, or phenotypic selection must be called "natural selection," one must make it absolutely clear precisely which restricted meaning one intends. In this book, I will use the general meaning of "natural selection" rather than the restricted meanings.

1.4. MODES OF SELECTION

Natural selection may affect populations in a number of different ways or modes. There are basically six classes of modes, each emphasizing different aspects of the process. They relate to differences in (1) trait mean, variance, and covariance; (2) number of equivalent phenotypes; (3) effects of other phenotypes; (4) habitat diversity and habitat choice; (5) levels of selection; and (6) mode of inheritance. These are independent, and several may happen simultaneously in the same population.

1.4.1. Mean, Variance, and Covariance

Natural selection is a process that affects the frequency distributions of heritable traits of a population. Traits may vary continuously or discontinuously (Figure 1.3). Continuously varying traits are often called quantitative (or morphometric) traits, and discontinuously varying traits are often called polymorphic

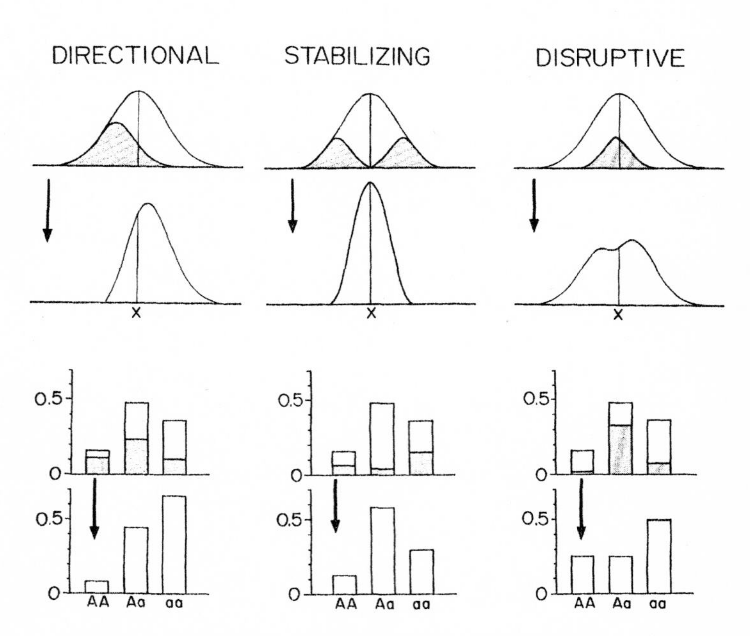

FIGURE 1.3. Three modes of selection for quantitative (upper row) and polymorphic (lower row) traits. In each case the vertical axis is the proportion of individuals, and the area under the curve, or the set of 3 bars, represents the total number of individuals. The individuals in the shaded portion are at a relative disadvantage compared to the individuals in the unshaded portion of the population. The arrows point to the offspring distribution after selection. X marks the mean of the distribution of quantitative trait values before selection, and the labels *AA*, *Aa*, and *aa* refer to three genotypes at a single polymorphic locus. In truncation selection (not shown) individuals above or below a threshold value x would be at an advantage, rather than have the more probabilistic relationships shown here.

traits. A third class of traits, quantitative threshold traits, has so far received virtually no attention in natural populations. These traits are phenotypically expressed in discrete classes, and so appear polymorphic. But they are inherited as though they were determined by an underlying quantitative variable with thresholds; the thresholds determine to which discontinuous phenotypic class an individual belongs. See Falconer (1981) for a fascinating discussion of this subject. In this book, the term "quantitative trait" will refer to quantitative threshold traits as well as ordinary continuous traits. In actual practice, the means and variances of the underlying continuous variables of threshold traits will have to be worked out by genetic analysis.

Natural selection affects these trait classes in somewhat similar ways, though their associated theories are quite different. A major difference is that quantitative trait theory explicitly includes the effects of environment on phenotype (Falconer 1981; Bulmer 1980), whereas polymorphic trait theory (Ewens 1979; Wallace 1981) assumes that, except for simple dominance, phenotypic variation is the same as genotypic variation. See Milkman (1982), Kimura and Crow (1978), Crow and Kimura (1979), and Lynch (1984) for more discussion of the relationships between selection of polymorphic and quantitative traits.

Simpson (1944) and Mather (1953) were the first to point out three ways in which natural selection can affect frequency distributions; this applies to both quantitative and polymorphic traits (Figure 1.3). In *directional* selection individuals toward one end of the distribution are favored. The mean will change, and the variance may decrease. In *stabilizing* selection intermediate individuals do better than the extremes; there is an intermediate optimum value. The variance will decrease, but the mean will not change unless there is a significant difference between the population mean and the mean selective value or optimum. In *disruptive* (or diversifying) selection extreme individuals do better than those with more intermediate characteristics. One possible reason for this is density-dependent or frequency-dependent selection; individuals with phenotypes closer to the mean will be more common and hence will be at a relative disadvantage. Another form of disruptive selection results when two different optimum values (or niches) are present and independent of the current trait distribution. In both forms of disruptive selection the variance will increase, but the mean will not change unless there is a significant difference between the population mean and the optimum or the mean of both optima. Note that these predictions only work if one form of selection affects the trait. It is quite possible for more than one mode to occur simultaneously; this will depend upon the trait distribution relative to the environment (Figure 1.4). Natural selection does not necessarily result only in a change in the mean and should therefore be described as well as defined

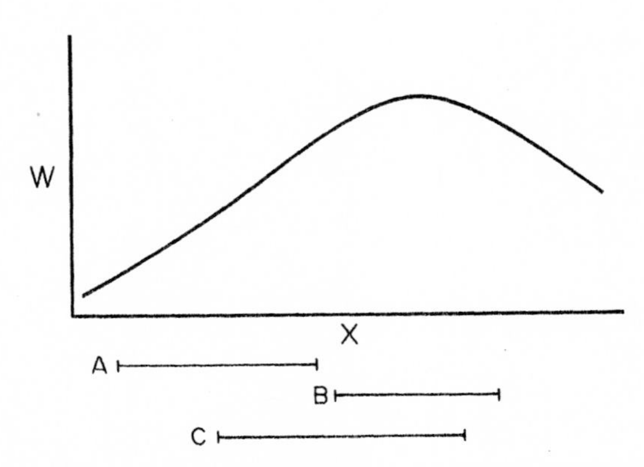

FIGURE 1.4. Selection mode and population variability. Consider a species that can vary in trait value X. For biophysical or physiological reasons, the fitness of an individual W varies with X as shown, no matter what part of the species geographic range we consider. A population showing the range of X values marked A will experience "pure" directional selection, while population B will experience "pure" stabilizing selection. Population C will experience a mixture of directional and stabilizing selection. The mode of selection depends upon the range of a population's variability as well as the fitness function.

in terms of the entire trait frequency distribution and the particular environmental conditions.

These predictions may also be affected if there are effects of selection for phenotypically correlated traits (Lande and Arnold 1983). If another trait is phenotypically correlated with the trait studied, and the other trait is subject to directional, stabilizing, or disruptive selection, then the apparent selection on the first trait may appear very different from the actual model of selection. The observed pattern of selection may be further modified by the presence of an additional mode, *correlational selection.* In correlational selection certain combinations of traits or alleles are favored at the expense of other combinations (Figure 1.5). This will result in patterns of gametic phase ("linkage") disequilibrium for suites of polymorphic traits (Ewens 1979), and patterns of phenotypic correlations for quantitative traits (Lande and Arnold 1983; Arnold et al. 1986). Correlational selection may not necessarily change the distribution of each trait considered by itself (Figure 1.5A), and so may not be detectable if only one trait is studied. Correlational selection may also give false evidence for other modes of selection, just as selection of certain traits can cause apparent directional or stabilizing selection in phenotypically correlated traits; we will return to this in Chapters 3 and 6. Once again, we must know as much as possible about the trait distributions and environmental parameters to understand natural selection.

Since natural selection does not necessarily result in a change in trait distributions among generations, nothing may be detectable if the population is at or near a stable equilibrium. An equilibrium can result from some form of stabilizing selection, a balance between directional selection and gene flow, or a balance between directional selection and a genetic bias such as meiotic drive. If a population is at or near equilibrium, then no change will be detected if the trait distributions are examined in successive generations at the same age classes, though there will be a consistent difference among the age classes. If there is directional selection, then differences will be seen among age classes and generations. For this reason the deductions from the three conditions for natural selection are stated in terms of differences among age classes, and the second deduction is conditional.

Natural selection may effect the distribution of trait values either as a smooth function of value, as in Figure 1.3, or with respect to a threshold. Truncation selection is a special form of directional selection in which individuals with trait values above (or below) a critical value survive or reproduce while

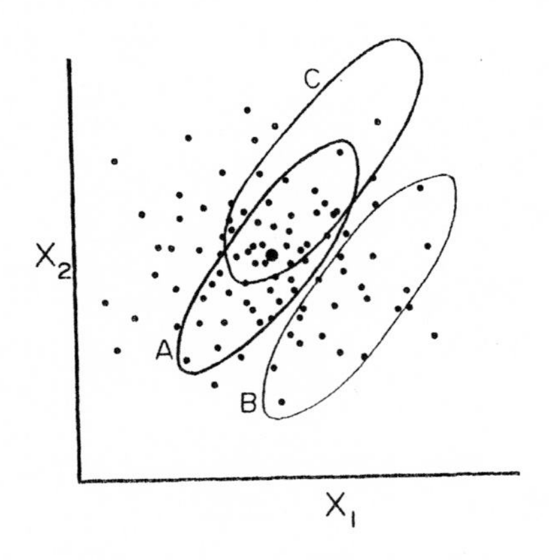

FIGURE 1.5. Correlational selection. Consider a species with two traits 1 and 2 which vary as shown: a single point represents an individual with trait values X_1 and X_2. The large dot is the population mean of both traits. The three ellipses *A*, *B*, and *C* indicate three possible results of selection, where the individuals within an ellipse are those that survived or bred successfully. *A*, "pure" correlational selection: here the means of X_1 and X_2 do not change during selection. *B*, correlational and directional selection: here directional selection strongly effects trait 1 and weakly effects trait 2. *C*, correlation and directional selection: here directional selection effects both traits equally. In all three cases the correlational selection favors similar combinations of trait values, and there is also some stabilizing selection. As with Figure 1.4, a complete knowledge of the distributions and environment is very important.

those on the other side of the threshold do not. Truncation selection is more efficient than gradual directional selection; the relationship between the two is discussed by Crow and Kimura (1979).

1.4.2. Number of Equivalent Phenotypes

In studies of natural selection, it is often assumed that all genotypes, phenotypes, or trait values are selectively different. But this is unrealistic. The distinctions we perceive between phenotypes may be different from how phenotypes are affected by natural selection. It is quite possible for a group of individuals of two or more phenotypes to be selectively neutral with respect to a particular environment, selectively advantageous relative to a second such group in the same environment, but disadvantageous to the second group in another environment (Endler 1978; Milkman 1982). In other words, there are multiple genetic "solutions" or outcomes to the same environmental factors or "problems" (Bock 1959; Gould and Lewontin 1979). I will refer to the selectively equivalent phenotypes within groups as "equivalent phenotypes." There may be no differences in condition *b* (fitness) or in the effects of natural selection among individuals within a group of equivalent phenotypes, but only among individuals of different groups.

Examples of equivalent phenotypes are known for variation in secondary compounds in plants, which affect palatability to herbivores; many different combinations of secondary compounds yield the same degree of unpalatability (Cook et al. 1971; Simons and Marten 1971; Sturgeon 1979). Other examples are found in seed color and survival (Brayton and Capon 1980), life-history patterns in *Corythucha* and *Gargaphia* lacebugs (Tallamy and Denno 1981), and color patterns in fishes (Endler 1978, 1980, 1983). For example, guppies (*Poecilia reticulata*) with many different combinations of genetically determined spots of particular sizes, colors, and brightnesses can be equally cryptic on a given background but conspicuous on another background. Another group of guppies with different phenotypes will be conspicuous on the first background, but cryptic on the second. There is nothing about selection that necessarily favors a *single* best genotype, phenotype, or phenotypic value, though it may happen briefly in a single locality.

1.4.3. Effects of Other Phenotypes

The mode of selection may or may not depend upon phenotypic composition of the population. In frequency-dependent selection, condition *b* (fitness differences) depends upon the relative frequency of similar phenotypes, and in density-dependent selection it depends on the relative as well as absolute numbers of phenotypes. (In some density-dependent models fitness may depend only on total numbers.) Wallace (1981) also makes a distinction between "hard" and "soft" selection. In "hard" selection successful phenotypes are better than an absolute criterion value; the presence of other phenotypes does not affect their relative success rate. Ground squirrels above a critical size may be the only ones to survive a severe winter, no matter what the size of their neighbors. This may result in the extirpation of some populations with very small mean sizes. On the other hand, in "soft" selection, success depends upon which other phenotypes are present, and it may depend upon relative or a combination of relative and absolute criteria. The larger squirrels in each population may have more offspring because they are better able to defend territories or to escape predators. Thus in summer the successful squirrels in populations of small squirrels will be smaller than equally successful squirrels in populations of large squirrels. As we can see, there is no simple one-to-one relationship between "hard" or "soft" selection and density-independent or density-dependent selection (for further discussion, see Wallace 1981, Walsh 1984, and references therein). Clearly, the dynamics and outcome of selection will depend upon the interactions among phenotypes as well as the interactions of phenotypes and their environment.

1.4.4. Habitat Diversity and Choice

There is no reason for the mode of natural selection to be the same throughout the geographic range of a species, and most species ranges contain many different habitats. Geographic variation in selection, along with gene flow, can cause geographic variation in gene frequencies and trait values, and maintain variation in a way that would appear impossible if only one locality were analyzed (reviewed in Endler 1977). Spatial variation in selection on a scale equal to or smaller than a deme or neighborhood can also maintain variation, though the conditions are more stringent for these "multiple-niche polymorphisms" (Ewens 1979; Maynard Smith and Hoekstra 1980; Hedrick 1983). If there is any tendency for different phenotypes to actively choose microhabitats in which they are favored, then the conditions are less stringent (Jones and Probert 1980; Maynard Smith and Hoekstra 1980). Though there is much theoretical work on the subject, virtually no work has been done to these ideas in natural populations. This would repay further study.

1.4.5. Levels of Selection

It is theoretically possible for selection to have different effects at levels other than the individual phenotype—for example on genes, genotypes, groups, populations, species, or even self-replicating molecules (Dawkins 1976, 1982; Sober 1984). Sober (1984) provides a superb discussion of the units of selection problem. In his discussion of the difference between "units" and "levels" of selection, Brandon (1982) presents an interesting method to help decide at what level (or levels) natural selection should be considered:

> Natural selection occurs at a given level if and only if:
> 1. Conditions *a*, *b*, and *c* [see section 1.1] hold at that level.

2. The expected fitness of those entities screen off the expected fitness of entities at every other level from that at the given level. (Rephrased from Brandon 1982.)

"Screening off" is a method of statistical inference, devised by Salmon and Reichenbach to attempt to assign primary cause in a complex cause-effect relationship (Salmon 1971). For the moment, we will define fitness as the differences expressed in condition *b*; this will be discussed further in Chapter 2. Following Brandon's discussion: If *A* renders *B* statistically irrelevant with respect to outcome *C*, but not vice versa, then *A* is a better causal explainer of *C* than is *B*; *A* screens off *B* from *C*. In other words:

$$P(C|A \cdot B) = P(C|A) \neq P(C|B). \qquad (1.1)$$

[$P(C|A \cdot B)$ is the probability of *C*, given *A* and *B*.] More remote causes are screened off by more immediate causes. Brandon's examples are quite revealing: (1) A sudden drop in atmospheric pressure (*A*) screens off a drop in barometer reading (*B*) from the the occurrence of a storm (*C*); (2) phenotypes (*A*) screen off genotypes (*B*) (and genes) from differential reproduction (*C*). "Tampering with the phenotype without changing the genotype can affect reproductive success (as my castrated cat testifies), while tampering with the genotype without changing the phenotype cannot affect reproductive success" (Brandon 1982, p. 317). The latter is another restatement of Mayr's (1963) argument that phenotypes, rather than genotypes or genes, are affected by natural selection. One major problem with this method is that it is not always clear that what seems to be the most proximal cause is indeed the primary cause; this is the fundamental problem of distinguishing cause-effect from correlation.[4]

[4] For further discussion, including a good one of group selection, see Williams 1966; Brandon 1982; and especially Sober 1981, 1984. A collection of important papers on the levels and units of the selection problem is reprinted in Brandon and Burian 1984.

There has been considerable interest in the possibility of group selection, or natural selection resulting from the differential proliferation or extinction of groups of individual organisms (Dawkins 1976; Wade 1978; Wilson 1983; Sober 1984). Nunney (1985) provides a very useful discussion of the various meanings of the term, and the conceptual difficulties are lucidly discussed by Sober (1984). Wade (1978) makes a fine review of the literature and has provided the only critical evidence for its existence—in the laboratory. The conditions for group selection are rather stringent (Wade 1978), suggesting that it may not be common in nature. The only evidence for group selection in natural populations is indirect and comes from successful prediction of sex ratios in the presence of local mate competition (summarized in Charnov 1982). In any case, there is no question that population structure affects the outcome and mode of natural selection (Wade 1978; Wilson 1983; Sober 1984; Nunney 1985). Another possibility is species selection (Stanley 1979; Gould 1982; Sober 1984). It is based upon the assumption—for which there is some evidence (Jablonski and Lutz 1983)—that certain kinds of species tend to speciate, and go extinct, more often than others. Since the evidence is weak or incomplete, group and species selection will not be discussed further in this book. See Sober (1984) for a thoughtful discussion.

1.4.6. Mode of Inheritance

The definition of natural selection does not necessarily apply only to strictly genetic traits. In social animals that transmit information from parents to offspring, or more generally between a group of parents and all their offspring, condition *c* (inheritance) is met. Given also conditions *a* (variation) and *b* (fitness differences), natural selection of cultural traits will result. Good examples include the spread of milk-bottle cap opening behavior in British *Parus* (Sherry and Galef 1984), and agriculture in man (Ammerman and Cavalli-Sforza 1984). Cultural selection may also have interesting side effects on "regular" genetic inheritance (Cavalli-Sforza and Feldman 1981; Ammerman and Cavalli-Sforza 1984). See Dawkins (1976, 1982), and especially Cavalli-Sforza and Feldman (1981), for reviews of cultural inheritance. Common-environment effects (footnote 2) may also yield apparently heritable effects for several generations, blurring the distinction between genetic and cultural inheritance. An example is found in red deer: accidental differences in a deer's food and conditions can have significant effects on the fitness of its children and grandchildren (Clutton-Brock et al. 1982).

1.5. SUMMARY

Natural selection is a process that results from biological differences among individuals, and which may give rise to cumulative genetic change or evolution, but does not guarantee it. The process is derived from a syllogism; given three conditions for natural selection (*a*, variation; *b*, fitness differences; *c*, inheritance), two conclusions necessarily follow: (1) differences in trait distributions among age classes or life history stages; and (2) if the population is not at equilibrium, a predictable difference among generations. By the nature of its definition, natural selection can be broken down into various components, either by separation into alternative (sexual and nonsexual selection) or sequential subprocesses (phenotypic selection and genetic response). These components are all parts of the same process, but restricted meanings of the term "natural selection" to one or more of its components are confusing when the restriction is not made explicit. Natural selection may occur in various modes, and the diversity of mode classifications represent different aspects of the process.

CHAPTER TWO

Philosophical Comments

> It is one of the maxims of the civil law that definitions are hazardous. Things modified by human understandings, subject to varieties of complication, and changeable as experience advances knowledge, or accident influences caprice, are scarcely to be included in any standing form of expression, because they are always suffering some alteration of their state.
>
> Samuel Johnson, *The Rambler*, 28 May 1751

Although it is a relatively simple concept, natural selection is associated with quite a few philosophical problems. Arguments about the existence and logical status of natural selection have crossed many fields, and have involved philosophers, biologists, phylogenetic systematists and others.[1] Disputes have resulted from incomplete understanding, including confusion between cause and effect, and historical accident. In addition, many problems are due to progressive modification of concepts without modification of terminology: the same words have completely different meanings to different people. We have already had an example of this in the discussion of restricted meanings of natural selection (Figure 1.2), but problems of nonstandard terminology are deeper than this. It is the purpose of this chapter to summarize briefly some of these problems, suggest some solutions, and offer some suggestions for standardization in order to

[1] *Philosophers:* Reed 1981; Hull 1980; Popper 1963, 1972, 1974, 1978; Mackie 1978; Ruse 1977; Caplan 1977; Sober 1984; Urbach 1978; Wasserman 1978, 1981; M. B. Williams 1970, 1973. *Biologists:* Ghiselin 1969; Lewontin 1974; Gould 1982; Gould and Lewontin 1979· Peters 1976; Castrodeza 1977; Stebbins 1977; Stern 1970; G. C. Williams 1966. *Phylogenetic systematists* ("cladists"): Eldredge and Cracraft 1980; Platnick 1977; Rosen 1978; Rosen and Buth 1980. *Others:* Macbeth 1971; Bethell 1976.

reduce confusion in the future. The discussion is not meant to be definitive, and is meant only to serve as a general guide to the problems.

2.1. NATURAL SELECTION AND TAUTOLOGY

Natural selection has been called a tautology by systematists, philosophers, and others.[2] This results from the mistaken use of "survival of the fittest" as a synonym for natural selection. This unfortunate phrase was first used by Herbert Spencer, borrowed by Darwin for later editions of *The Origin of Species* (1869), and amplified by biologists such as Haldane (1935) and Waddington (1960, p. 385). A tautology is a statement that is true by a definition contained within the statement. Natural selection, as defined in Chapter 1, is not a tautology but a syllogism in which two new conclusions are deduced from the premises. "Survival of the fittest" is certainly a tautology; but it completely misses the point of natural selection and confuses cause and effect. In addition, survival is only one component of the process.[3] This early dual mistake was quickly rectified (Mayr 1942; Simpson 1949) or was never made to begin with (Fisher 1930; Wright 1931). Unfortunately, "survival of the fittest" was picked up by philosophers such as Popper, by non-biologists (Macbeth 1971; Bethell 1976), and then by cladists,[4] who rightly criticized that phrase but then threw out the entire concept of natural selection. It is only recently that philosophers,[5] including Popper (1978), have realized that natural selection is something more than "survival of the fittest." For

[2] See Popper 1963–1974; Peters 1976; Platnick 1977; Rosen 1978; Rosen and Buth 1980; Cox 1981; Pearson 1981; Macbeth 1971; Bethell 1976.

[3] Chapter 1; see also Mackie 1978; Mayr 1942; Reed 1981; Ruse 1971; Sober 1984; Urbach 1978; Wassermann 1978, 1981; G. C. Williams 1966; M. B. Williams 1970, 1973.

[4] See note 1.

[5] Mackie 1978; Reed 1981; Ruse 1977; Caplan 1977; Sober 1984; Urbach 1978; Wassermann 1978, 1981; M. B. Williams 1973.

an excellent review and discussion, see Wassermann (1978) and Sober (1984). If one must use a catch phrase to describe natural selection, Oscar Wilde's is much better: "Nothing succeeds like excess." This is very close to Fisher (1930), even to the use of "excess."

2.2. FORCE, ACTION, AND INTENSITY

It is common in the literature, and in textbooks, to say that natural selection "acts" on organisms or gene pools, and that it is a "force" in evolution. This usage goes back to Darwin (1859, p. 90), but it does not seem to be generally appreciated that he used "acts" for convenience.

> The term "natural selection" is in some respects a bad one as it seems to imply conscious choice; but this will be disregarded after a little familiarity. No one objects to chemists speaking of "elective affinity"; and certainly an acid has no more choice in combining with a base, than the conditions of life have in determining whether or not a new form be selected or preserved. . . . For brevity sake I sometimes speak of natural selection as an intelligent power:—in the same way as astronomers speak of the attraction of gravity as ruling the movements of the planets. . . . I have, also, often personified the word Nature; for I have found it difficult to avoid this ambiguity; but I mean by nature only the aggregate action and product of many natural laws,—and by laws only the ascertained sequence of events. (Darwin 1875, vol. 1, pp. 6–7)

Unfortunately, others have not been so careful, and this has often resulted in tautologies. Natural selection no more "acts" on organisms than erosion "acts" on a hillside (Ghiselin 1969, 1981). It is a *result* of heritable biological differences among individuals, just as erosion is a result of variation in resistance to weathering and running water. Admittedly, it is convenient, almost poetic shorthand, to say that natural selection (or erosion) "acts" and is a "force," but this misses the point, misrepresents the process, and can easily give rise to tautologies as well as problems concerning the "units on which selection acts" (Ghiselin 1969, 1981; Lewontin 1970; Dawkins 1976; Hull 1980). Ghiselin says it very well:

> We are not dealing with a change in substance of a straightforward kind, but with a process that influences relations. A simple analogy occurs with the removal of clothing. I remove some garments. I act. Undressing, not being the agent, does not "act" at all in that sense. My clothes are acted upon. I am affected—not by my acting on myself *simpliciter*, but through a change in the relationships between myself and my garments. Does it make any sense to say that a shirt is a "unit of undressing"? If there is any such unit here, I would much prefer to have it the removal itself. But then again, the most fundamental unit is perhaps the thread, which, of course, is not removed until the weaver's art has expressed it phenotypically. Some people say that selection is "purely negative" and does not really explain evolution: Does undressing fail to explain nudity, or erosion the configuration of the landscape? (Ghiselin 1981, p. 282)

The problems in considering "units" and "levels" of selection[6] are clarified by this viewpoint. Since natural selection does not "act," it is meaningless to speak of a "unit" of selection, except possibly as an entity whose *relationships* with other such entities are *affected* by selection. This entity is the genotype and not the gene because gene arrangements can be affected by selection even when gene frequencies do not change (Dobzhansky 1970; Sober 1984). It is not very helpful to think of a mountain or a drainage system as a "unit of erosion"—the cause of erosion

[6] See Lewontin 1970, 1974; Dawkins 1976; Hull 1980; Sober 1981, 1985; Whimsatt 1981; Brandon 1982.

is variation of weathering resistance among rocks and soils at various geographic scales, in conjunction with fluctuations in temperature and moisture. As for genes, genotypes, phenotypes, groups, populations, and species, we can consider the processes of erosion at various scales or levels; but what is most important are the biological differences that result in natural selection. Another way to say this is that there is a significant difference between "units" and "levels" of selection (Brandon 1982). In part, "units of selection" may be artifacts of thinking that natural selection "acts." (See Sober 1984 for a lucid discussion of "units" of selection in which there is no need to distinguish "units" from "levels.")

Natural selection is, in a sense, a "force" that can "act" on gene frequencies or trait value distributions, and this is the most common sense in which "act" and "force" are used. But this is only a vague and most improper analogy with physics.

First, if natural selection is a "force," what is it acting *on*? (A force is meaningless without an object.) If natural selection were a force, it should be possible to decompose it into a mass and an acceleration. In this case "acceleration" is phenotypic selection, but what is the "mass"? The "mass" could be a frequency distribution or the genetic system (condition *c*, inheritance), but this is tantamount to assuming that natural selection applies only to groups. Natural selection arises from biological differences among individuals (condition *b*, fitness differences); therefore to make a proper analogy, the "mass" is the genetic composition of an individual. This is reasonable because it also allows mutation to be a "force." But the "mass" in the physicist's $F = ma$ is a *class* of objects with defining properties and not an individual (Ghiselin 1981), so the analogy either breaks down or restricts natural selection to group selection. The use of "force" in describing natural selection is an instance of confusion between the change in an object and the changes in *relationships* among objects, and can also be an artifact of typological (class) thinking.

Second, and more important, physicists use "force" when they do not know why objects are accelerated, for example by gravity or magnetism. But this is not necessary in natural selection, where we can find out *why* condition *b* is true, or why frequency distributions change. Natural selection is not like a physical system of forces moving objects; it is more like a system of chemical reactions, which consist of a population of molecules with different properties. This is a good analogy because a reaction may go to completion or reach a dynamic equilibrium. Chemists do not talk about a mysterious "force" changing the frequencies of the reactants, but about the properties of the reactants that lead to the process. The use of "force" confuses cause and effect. Like phlogiston in thermodynamics, "force" can be an empirical description of the process of natural selection, but it has nothing to do with its cause or its mechanics.

There is only one usage of "force" that is not misleading, and that is "force" in the sense of a *causal influence*. Certainly a causal influence can act. But like the physicists' "force" this does not say *why* the cause acts, and gives us no new biological insight into the process of evolution; it takes the emphasis away from the primary causes, namely conditions *a*, *b*, and *c*. But if we mean "force" merely as a cause, it is clearer to use "cause" directly, rather than burying it in the force analogy.

It is common to talk about the "intensity" of selection, but this is part of the same confusion about "force." There is no "intensity" of chemical reactions (or erosion), but there are rate constants or coefficients that describe and predict the dynamics of the reaction. In the same way, it is improper to speak of an "intensity" of natural selection, but quite reasonable to speak of its rate or the magnitude of its rate coefficient.

To reiterate: natural selection is a process resulting from the interactions between variable organisms and environment, and not a mysterious "force" that "acts." The use of "acts," "force," and "intensity" has the same status as phlogiston in thermodynamics; it is descriptive but misrepresents the process, confuses cause and effect, and focuses attention away from causes. The important and interesting questions in studies of natural selection are: (1) What are the biological reasons that some trait values have higher fitness than others or what are the biological reasons for natural selection? (2) Given that there is fitness variation, what are the evolutionary dynamics and equilibrium configuration (if any) of the trait? Natural selection cannot "generate genetic changes within populations," it is a *process, resulting* from heritable biological differences among individuals, which can lead to genetic change in populations and species.

2.3. FITNESS AND ADAPTATION

Fitness is an estimate and a predictor of the rate of natural selection, and is therefore necessary for a good understanding of natural selection. If "fitness" is used in a careless fashion, it can lead to tautology, since by definition the most fit survive or otherwise do better than the less fit. As mentioned earlier, this is one reason for a mistaken criticism of the concept of natural selection. This criticism misses the point entirely. Given the process of natural selection, some genotypes may be more successful, and this can be demonstrated as a result of the analysis of a natural population. We *define* these to be more fit, or to have greater fitness; fitness is a convenient word that summarizes either the presence of condition *b* for natural selection (fitness differences), or the result of ongoing natural selection. Fitness is not an explanation, it is only a *description*. It quantifies the biological properties that lead to shifts in genotype distributions; it does not cause the shifts. For an excellent discussion of the relationships among fitness, natural selection, and evolution, see Sober (1984).

There are many different definitions and measures of fitness, each trying to capture its essence. The development of these measures has often proceeded independently in different fields, leading to duplication under different names, and emphasis on different aspects of fitness under the same name.[7]

In trying to sort out the literature, Cooper (1984) made the useful distinction between fundamental (or direct) and derived (or indirect) measures of fitness. A fundamental measure of a variable measures some property of the variable directly, while a derived measure depends upon some property derived from the original variable, and it may or may not be directly correlated wth the fundamental value. Derived measures may be incomplete measures, and they assume that everything else is being held constant (Cooper 1984). A measure of fitness that includes only survivorship would not be very accurate unless it was known that all other components of fitness (mating ability, fecundities, etc.) were equal (Fisher 1930). In the next two sections, I will discuss Cooper's ideas on fitness, along with five other concepts, and place them in perspective.

2.3.1. *Expected Time to Extinction*

Cooper (1984) argues that the most fundamental measure of fitness is the expected time to extinction (ETE) of alleles, and that all other measures are derivatives of it. A similar argument was made by Thoday (1953, 1958). Cooper discusses eight ways of measuring fitness; he shows that they are either derivative measures of ETE, or that ETE is at least as good a measure of fitness. They are: (1) finite rate of increase (W, R, or λ, average number of offspring per individual) or instantaneous rate of increase (r), assuming constant absolute growth rates; (2) as in (1), but assuming constant underlying growth rates; (3) as in (1), but assuming constant relative growth rates (the population genetics model); (4) as in (1), but assuming stochastic regulation that is independent of genotype, (5) as in (2), using geometric mean growth rate, and assuming stochastic variation

[7] See Dobzhansky 1956, 1968a,b; Cooper 1984; Dawkins 1976, 1982; Ewens 1979; Falconer 1981; Kempthorne and Pollak 1970; Prout 1965–1971; Stern 1970; Sober 1984; Thoday 1953, 1958; Wallace 1981.

in growth rates (this method and model were also considered by Haldane and Jayakar 1963; Gillespie 1973; and Karlin and Lieberman 1974); (6), as in (4), but with genotype-dependent stochastic regulation; (7) probability of eventual extinction; and (8) probability of extinction by time t.

Models (7) and (8) are logically flawed (Cooper 1984). Since everything must become extinct sometime, the probability of eventual extinction of all genotypes is always 1.0. The probability of going extinct at a specified time t is dependent upon an arbitrary t. Consider two genotypes A and B: if t is too small, then neither A nor B will go extinct, and if t is too large, both will have disappeared. For both (7) and (8) the ETE is the only sensible measure. For the deterministic models (1), (2), and (3), the ETE is as good as the other measures. But for models with stochastic parameter variation, ETE is not only as good as but is sometimes better than the alternate measure.

Consider the stochastic models (4, 5, 6). (4): Let W_A and W_B be the absolute or relative growth rates of A and B, respectively. Whenever the total population becomes larger than K, then the probability that an individual of either genotype survives and reproduces is L. Thus, on average the population and each genotype are reduced to a fraction L of what they were when the population first exceeded K. For any particular population of A and B, if $W_A > W_B$, then there is no guarantee that A will exist longer than B; by chance A may go extinct first. It is only true that *on average* the time to extinction of A will be longer than B; thus ETE is a better representation of the outcome of natural selection than, say, $(W_A - W_B)/W_A$. Consider (5): Let the W's vary at random with means W_A and W_B, and their geometric means in time can be used to estimate fitness. (For detailed discussions, see Kimura 1954; Haldane and Jayakar 1963; Gillespie 1973; Karlin and Lieberman 1974.) Once again, for any particular population there is no necessary relationship between W and its time to extinction. For example, if $W_A = W_B$, but the variance of W_B is greater, then A may last longer than B, even though their geometric mean fitnesses are equal. Gillespie (1977) gave a similar result for the effect of variance in offspring number. Once again ETE is a better measure of fitness. Consider (6): A and B are regulated by L_A and L_B, respectively. If the L's vary stochastically, once again ETE is a better measure of fitness, though L's are fine if the model is deterministic. Note that these models are for selection of haploid organisms; recombination can seriously affect the conclusions.

All methods, except for the expected time to extinction, depend upon particular specific models, and the ETE works in all, and ETE works better (or is the only one that works) where there is stochastic variation in fitness (Cooper 1984). For example, for the deterministic versions of (4), (5), and (6), different measures of fitness must be used (W's or L's or both), but ETE can be used for all. The ETE appears to be a fundamental measure of fitness. A genotype or character value with a longer expected time to extinction is one that is favored by natural selection.

One has to be careful; this statement can be misread as the tautological definition of natural selection. One genotype has a larger ETE and, by definition, is favored by selection. It is favored not because it has a larger ETE, but because its biological properties *result* in a larger ETE. The ETE is an indication of the result of the process of selection or evolution, not its cause.

There are two serious practical and theoretical problems in using ETE's. First, we are not omniscient and cannot estimate the ETE directly because we cannot replicate a particular population's history a large number of times to calculate its ETE. Suppose that the ETE's of A and B were equal (or that the means and variances of their W's and L's were equal). By accident B may go extinct sooner than A, even though A and B were biologically identical; but by the single estimate of the actual time to extinction B is less "fit." Suppose that the ETE of A is greater than B, but by accident (such as a major climatic change) A went extinct first. B would then be regarded as more "fit," even though B would have gone extinct sooner if the rare catastrophic event had not occurred. The problem of rare catastrophic events also affects other measures of fitness, with less serious effects. (For a discussion of the probabilistic nature of fitness, see Sober 1984.) Still another serious measurement problem exists because we usually do not live long enough to witness the extinction of an allele. Thus, although the ETE seems a very good measure of fitness, especially where there is stochastic selection, it is not very practical. It is impractical in still another way: it cannot be used in predicting the dynamics of natural selection or the equilibrium distribution of genotypes; at best it can be used only to say something about the rate of evolution. What is clearly needed is a careful mathematical treatment of the effects of random variation in various components of fitness on the rate of evolution, as has been done for population dynamics and regulation (May 1973). The papers by Gillespie (1973, 1977) and Karlin and Lieberman (1974) are interesting pioneering efforts.

A second serious problem with ETE is that, although it may predict the relative success of alleles, special models are required to show that it will be increased by natural selection. Using the results of Lewontin and Cohen (1969) we can show the following (Charnov, pers. comm. 1984): If a population is growing exponentially (a function of r), and if the geometric mean of r (averaged over time) is greater than 1, then the population cannot go extinct. If the geometric mean of r is less than 1, then it is certain to go extinct. If $r < 1$, then the time to extinction is proportional to $2v/m$, where m is the geometric mean of r, and v is the variance of $\ln(r_t)$ (r_t is the intrinsic rate of increase at time t). The result is almost independent of initial population size and criterion of extinction. Therefore, if natural selection tends to increase the time to extinction, it must increase the coefficient of variation of r. The conditions that favor an increase in v/m are discussed in detail by Gillespie (1977), Cooper and Kaplan (1982), and Kaplan and Cooper (1984); but in other equally realistic models, there appears to be no tendency to increase v/m. Therefore ETE will not necessarily increase with time, and it may only be used to suggest which allele may be more successful. For further discussion of problems arising out of chance events see Crandall et al. (1985).

ETE is not an ideal measure of fitness. It would be much better if the measure were intimately bound up in the causes of natural selection, rather than merely describing its result; we would like to know how and why a particular allele or genotype happens to have a longer expected time to extinction. In that sense, the ETE is a derived measure. An ideal fundamental measure of fitness would contain the notion of the mean lifetime contribution of a given genotype to the next generation, as well as its variance.

2.3.2. *Five Meanings of Fitness*

The discussion of the merits and faults of ETE illustrates how the quality of a fitness definition depends upon its purpose and frame of reference. For example, ETE is concerned with the success of alleles over many generations, rather than the success of individuals in one generation. Dawkins (1982) provides an excellent discussion of the purposes of definitions and their frames of reference with respect to the levels of selection (genes, genotypes, individuals, groups). But as the ETE example suggests, there is another dimension of complexity. Various definitions of fitness try to get at two different phenomena: (1) the tendency for some phenotypes (quantitative and polymorphic) to have properties (condition *b* for natural selection), which result in their being differentially represented among age classes or generations; and (2) the tendency for some genotypes to replace others over many generations. This is a reflection of the difference between natural selection and evolution (Table 1.1). A lack of distinction among these phenomena has led to considerable confusion in the literature, as well as to some unwarranted criticisms of natural selection.

A major problem in distinguishing various concepts of fitness in the literature is that there is at present no consensus on their definition, and, worse, the names exchange meanings as one goes from author to author. To make matters even more complicated, fitness is often confused with adaptation (for example, Stern 1970). The concepts of fitness and adaptation are related, but are not identical (Dobzhansky 1968a,b; Dunbar 1982; Sober 1984). In addition, there are at least five different concepts that are covered by the terms "fitness" and "adaptation." There is utter disagreement even about which concept applies to genes, individuals, populations, or species (for example Bock 1980; Dunbar 1982). In the hope that it will encourage standardization of terminology and thereby reduce confusion, I offer the five definitions in Table 2.1. Most of them are derived from the very clear discussion by Dobzhansky (1968a,b). As an illustration of the confusion in the literature, each of these five concepts has been called "fitness" by one or more authors (reviewed in Dobzhansky 1968a and Stern 1970).

FITNESS. *Fitness* is the degree of demographic difference among phenotypes, or a measure of the degree of the relationships in condition *b* for natural selection. This is the meaning of "fitness" that will be used in the rest of the book. It can be measured by the average lifetime contribution to the breeding population, by a carrier of a phenotype or of a class of phenotypes, relative to the contributions of other phenotypes. This is also known as "Darwinian fitness," "relative fitness," and sometimes "selective value," and is the standard fitness of population and quantitative genetics (Ewens 1979; Falconer 1981; Hedrick 1983). It is also sometimes known as "adaptive value," but this encourages confusion with adaptation. "Selection coefficient" and "selection differential" are algebraically related measures for polymorphic and quantitative traits, respectively (see Chapter 6 and Milkman 1982). Fitness is a within-generation measure of the process of natural selection, and thus applies

TABLE 2.1. Concepts that have been called "fitness" by various authors (partly after Dobzhansky 1968a,b)

1. *Fitness:* The degree to which condition *b* for natural selection is true. Measured by the average contribution to the breeding population by a phenotype, or of a class of phenotypes, relative to the contributions of other phenotypes. Also known as "Darwinian fitness," "relative fitness," and "selective value." "Selection coefficient" and "selection differential" are algebraically related to fitness.
2. *Rate Coefficient:* The rate at which the process of natural selection proceeds. Measured by the average contribution to the gene pool of the following generation, by the carriers of a genotype, or by a class of genotypes, relative to the contributions of other genotypes. Very similar to fitness, but also includes the genetic response (condition *c*).
3. *Adaptedness:* The degree to which an organism is able to live and reproduce in a given set of environments; the state of being adapted. Measured by the average absolute contribution to the breeding population by a phenotype or a class of phenotypes. Also known as "absolute fitness." Is also applied to species, where it is known as the "Malthusian parameter," r_m, or R.
4. *Adaptability:* The degree to which an organism or species can remain or become adapted to a wide range of environments by physiological or genetic means. The reverse of specialization.
5. *Durability:* Probability that a carrier of an allele or genotype, a class of genotypes, or a species will leave descendants after a given long period of time. This is best expressed for alleles, genotypes, or species by the expected time to extinction.

only to phenotypic selection. It must be transformed through the genetic system (genetic response) to yield the rate of natural selection.

THE NATURAL SELECTION RATE COEFFICIENT. The *natural selection rate coefficient* determines the characteristic rate at which natural selection proceeds and is the only "fitness" that is directly analogous to rate coefficients (K) in chemical reactions. It can be measured by the average lifetime contribution to the gene pool of the following generation, by the carriers of a genotype, or a class of genotypes, relative to the contributions of other genotypes. This is Dobzhansky's (1968a) definition of "fitness," but it is quite different from the "fitness" he and other population geneticists actually use in practice (see previous paragraph). The major difference is that this is a *between-generation* measure, and therefore includes the effects of genetics (condition c, inheritance). Confusion between rate coefficients and fitness results in false frequency-dependence in fitness estimates (Prout 1965). The rate coefficient is for the whole process of natural selection, while fitness applies only to one component—phenotypic selection.

Note that the rate coefficient is different from the absolute rate of natural selection (as in chemical reactions). Absolute rate is the difference in gene frequency (Δp) or trait value (ΔX) between generations. To illustrate the difference, consider a single polymorphic locus with two alleles A and a, in a system that meets all of the Hardy-Weinberg assumptions (Chapter 3, section 3.4.1), except that there is selection against recessives. The genotypes and fitnesses are therefore AA: 1, Aa: 1, and aa: $1 - S$; S is the selection coefficient. Let p be allele A frequency and $q = (1 - p)$ be allele a frequency, yielding genotype frequencies in zygotes of p^2, $2pq$, and q^2. After selection the new gene frequency is $p' = p/(1 - Sq^2)$. This yields an absolute rate of $\Delta p = p' - p = Spq^2/(1 - Sq^2)$. The rate coefficients can be measured at the zygote stage at both generations as follows. The new zygote frequencies in terms of the *old* gene frequency and the selection coefficient S are:

$$AA\colon \frac{p^2}{(1 - Sq^2)^2} \qquad Aa\colon \frac{2pq(1 - Sq)}{(1 - Sq^2)^2} \qquad aa\colon \frac{q^2(1 - Sq)^2}{(1 - Sq^2)^2}.$$

These are divided by the old zygote frequencies, p^2, $2pq$, and q^2, respectively, to yield ratios analogous to absolute fitnesses:

$$AA\colon \frac{1}{(1 - Sq^2)^2} \qquad Aa\colon \frac{(1 - Sq)}{(1 - Sq^2)^2} \qquad aa\colon \frac{(1 - Sq)^2}{(1 - Sq^2)^2}.$$

Analogous to relative fitness calculation, the heterozygote ratio is then divided into the homozygote ratios to yield the rate coefficients:

$$AA: \frac{1}{(1 - Sq)} \qquad aa: (1 - Sq).$$

Note that the two coefficients are reciprocal, and both are measures of the characteristic rate of natural selection for a particular genotype. (In this case the rates are similar to the ratio of gene frequencies, but this is not true for other modes of selection.) Note also that they are frequency-dependent, even though condition *b* for natural selection (S) is constant; this is one of the problems which arise from confounding rate coefficients and fitnesses (Prout 1965–1971a,b). The situation is similar in chemical reactions.

ADAPTEDNESS AND ADAPTATION. *Adaptedness* is the degree to which an organism is able to live and reproduce in a given set of environments: the state of being adapted (Dobzhansky 1968a,b). *Adaptation* is the process of becoming adapted or more adapted (ibid.). Unfortunately, adaptation is also used in the sense of an adaptive trait (Lewontin 1978), confounding the end product with the process (see also Dunbar 1982). An *adaptive trait* is "an aspect of the developmental pattern of the organism which enables or enhances the probability of that organism surviving and reproducing" (Dobzhansky 1956, 1968a). There are problems in defining precisely what adaptedness is so that it can be measured (Dobzhansky 1956, 1968a,b; Stern 1970; Lewontin 1978; Dunbar 1982). One solution is to define it in the sense of absolute (rather than relative) fitness (Table 2.1). In this case it can be measured by the average absolute lifetime contribution to the breeding population by a phenotype or a class of phenotypes. It thus becomes intimately related to the actual (R) or intrinsic (r_m) rate of increase, or "Malthusian parameter," and these have actually been used as measures of fitness for populations and species, though there are some problems (Fisher 1930; Dobzhansky 1968a,b; Dunbar 1982). Adaptedness has also been defined as the *mean* absolute fitness (Sober 1984). There are some especially difficult problems in calculating absolute fitness when there are overlapping generations, when there is strong density-dependent selection, and when randomness in selection is important (Charlesworth 1980; Crandall et al. 1985). Much more work has to be done on methods to measure absolute and relative fitness under natural conditions.

The distinction between fitness and adaptedness is a useful one because the one does not necessarily imply or give rise to the other (Dobzhansky 1968a,b; Dunbar 1982; Sober 1984); a phenotype with high adaptedness may not have high fitness, and vice versa. This is obvious when we consider them as relative and absolute fitness respectively (Table 2.1). Natural selection results from *differences* in adaptedness (condition *b* for natural selection; see Chapter 1). If we were to regard survival as the criterion of adaptedness (as does Bock 1980), then a gene that increases longevity at the expense of fecundity will have high adaptedness but low fitness. This can be avoided by including reproduction as well as survivorship, as in the "Malthusian parameter" (r_m). However, this ignores the effects of density dependence; one phenotype may always have a greater r_m than a second, but the second phenotype may have a greater relative fitness at high density. This is part of the classical problem of measuring adaptedness and competitive abilities, as well as "r and K selection" in ecology (for discussion see Williams 1966; Southwood 1981; and Harvey et al. 1983). Even if there is no density-dependent regulation of numbers, fitness can still be decoupled from adaptedness. Consider two different populations that are initially homozygous for allele *A* at a locus. In the first population a mutant occurs to allele *B*, which is advantageous and increases in frequency but does not affect the total population size. In the second population, a mutant occurs to allele *C*, which increases in numbers because, say, *AC* and *CC* genotypes use resources more efficiently than *AA*; consequently the total population grows. In either population, both fitness and adaptedness of *AA* are smaller than those of the mutant genotypes. But it is possible for *AA* to have the same fitness in both populations, even though the adaptedness (r_m) of *C* genotypes is larger than that of *B*.[8]

Adaptedness and adaptation have also been used to describe traits with respect to functional design; the degree of adaptedness is measured by how close the trait is to, say, an engineering ideal. This is decoupled from fitness because the advantages of a trait being perfectly adapted to a particular function may be more than offset by other disadvantages (Dobzhansky 1968a,b; Lewontin 1978). The converse is also possible. Willows, cottonwoods, and many other riparian trees have wood that splits easily, so in the engineering sense they are poorly adapted, at least for structural integrity. However, the frequent loss of branches and limbs is an important method of vegetative reproduction: these trees sprout easily from fragments, which is an advantage for dispersal along streams and for increasing the probability of survival if the "parent" tree becomes undercut by the stream. Poor engineering adaptedness is in this case associated with high fitness. This illustrates how the functional design or engineering definition is inferior to the absolute fitness definition of adaptedness; riparian trees with more frequent limb loss (and the ability to sprout) will have a larger Malthusian parameter than those that do not fragment. This decoupling results because the design criteria only consider one function of the trait, rather than all its effects on the whole organism. It is always safer to use adaptedness in the sense of

[8] For further discussion, see Charlesworth 1971, 1980; Clarke 1972; de Jong 1983; Roughgarden 1971. A more general discussion is found in Sober 1984 and Crandall et al. 1985.

absolute fitness rather than functional design, since we rarely know enough about the ecology, genetics, and development of organisms to know all ramifications of a trait.[9]

Dunbar (1982) shows that it is a tautology to use relative fitness to measure adaptation because variation in adaptation gives rise to differences in relative fitness (condition *b* for natural selection). If we consider only mortality selection, this is merely a restatement of the "survival of the fittest" tautology. We can avoid this by measuring adaptedness as absolute fitness; the distinction between fitness and adaptedness then becomes one of relative versus absolute fitness (Table 2.1). This considerably clarifies Dunbar's (1982) discussion of the interdependence of the two concepts. However, this does bring up a fundamental problem: "There is no necessary reason why variants should leave different numbers of offspring other than the fact that they do so" (Dunbar 1982, p. 10). Condition *b* for natural selection is purely a *description* of the properties of the variants; it does not explain *why* the differences are present. To know why condition *b* is true we must know much about the biology and ecology of the organism and the genetics, development, form, and function of the traits of interest (Bock 1980; Dunbar 1982). These should yield the proximal reasons for the conditions of natural selection.

There is an additional problem: the proximal reasons do not explain, nor are they necessarily related to, the ultimate (fundamental) reasons for either condition *b* or adaptation. The process of adaptation is often taken to mean not only the increase in frequency of more adapted forms, but also the cumulative changes in adaptation and condition *b*. Thus a second meaning of adaptation is the evolution of fitness. But natural selection does not change the mean adaptedness (absolute fitness) of a population; it changes the mean *relative* fitness. A trait can be an adaptation

[9] Useful discussions, though only with respect to genetics and development, are found in Lewontin 1974, 1978, and Gould and Lewontin 1979. A general philosophical discussion is found in Sober 1984.

without making an improvement in adaptedness, and vice versa (see Sober 1984). The question of the evolutionary origin of the conditions for natural selection is distinct from the current running of the process. Natural selection is not an explanation for adaptation; it only explains why and how relatively better adaptations can increase in frequency. The answer to *why* a given adaptation is superior to an alternative one, or why condition *b* for natural selection occurs, is a question of the mechanism of origin of new variants, the history of their origin and spread, the biological and ecological reasons for conditions *a–c* for the variants, and the conditions themselves. This is a question of evolution; natural selection is only part of the story. To say that a new adaptation necessarily arose through natural selection is an incomplete description, a tautology, and a misrepresentation of natural selection, adaptation, and evolution. Natural selection addresses the problem of the *spread* of new variants or new adaptations, not their *origin*. I will return to this in Chapter 8.

Simpson (1944, 1949), Gans (1974), and Gould and Vrba (1982) were sufficiently concerned about the distinction between origin and spread of variants that they coined the words "preadaptation" (Simpson), "protoadaptation" (Gans), and "exaptation" (Gould and Vrba). These concepts are very similar and serve to emphasize the origin of a new trait value, or new function for an old trait. They also emphasize that a new function of an old trait may arise purely by chance, and have nothing to do with natural selection (if any) of the old trait. Of course, once the new function is achieved, natural selection associated with the new function will affect the trait distribution. Williams (1966) and Gould and Vrba (1982) would like to limit the term "adaptation" to traits that arose by chance or other factors, and were later affected by natural selection. Therefore "exaptation" applies to a trait with a new function so long as it is not affected by natural selection; it soon becomes an "adaptation". However, it is quite possible for a trait to remain an "exaptation" if it is not affected by natural selection. "Preadaptation" and "protoadaptation" are similar to "exaptation," but emphasize what *might* happen if a particular kind of natural selection takes effect; "exaptation" is a static description, while "protoadaptation" and "preadaptation" look to the possible. There is obviously some hair-splitting here, but the differences emphasize different histories. History also distinguishes them from fitness; as Sober (1984) suggests, "adaptation" is retrospective in that it implies something about the history of a trait, while "fitness" predicts the trait's future. Note that this usage of "adaptation" is only common within evolutionary biology; in other subjects the term merely indicates that a trait functions well for a particular purpose and implies nothing about natural selection. The multiplicity of meanings is unfortunate; one should always make clear which meaning is intended when using the word "adaptation."

The distinction between fitness and adaptedness is also useful when considering species, as shown by this example:

> In some species of trees, e.g., the California redwood, an individual tree may stand and produce viable seed for many centuries. Moreover, since the redwoods are capable of stump-sprouting, an individual tree may live indefinitely, as long as the external environment remains propitious. And yet, the individual near-immortality does not guarantee a perdurability of the species; in fact the California redwood is a relict species in danger of extinction. By contrast, certain insect species in which the individual is very short-lived seem to be thriving. (Dobzhansky 1968a)

Thus redwoods may have high adaptedness but low probability of survival relative to other species. (Note that high adaptedness includes reproduction in this example, as it should.) "Fitness" has not quite the correct meaning here, hence the next two definitions.

ADAPTABILITY. *Adaptability* is the degree to which an organism or species can remain or become adapted to a wide range of environments by physiological or genetic means (after Dobzhansky 1968a,b). The problem with the redwoods is that, although they are well adapted to their particular environment, their requirements are so narrow that they may not survive an environmental change. Adaptability is in a sense the opposite of specialization, and consists of evolutionary (genetic) plasticity as well as phenotypic plasticity. Note that the previous three definitions depended upon conditions in a particular environment, while adaptability depends upon many environments as well as the genetic and development properties of the species. Adaptability is a measure of actual or potential success given temporal and spatial variation, while fitness, rate coefficients, and adaptedness are measures of success only in the set of conditions in which they are measured.

DURABILITY. *Durability* is the probability that a carrier of an allele or genotype, a class of genotypes, or a species will leave descendants after a given long period of time (after Thoday 1953, 1958, and Dobzhansky 1968a,b). This is best expressed as the expected time to extinction (ETE), as discussed earlier. Durability is a function of the previous four concepts, with the vicissitudes of environmental change being at least as important as biological differences (condition *b* for natural selection) within a particular set of conditions. Unlike the previous concepts, this one cannot apply to individuals, though it applies to genes or species.

A related definition is one used by those who work with optimization models: the quantity that natural selection maximizes at equilibrium. The phenotype with the "evolutionary stable strategy" (Maynard Smith 1978) may also have greatest durability. But these two meanings really hold only for constant environmental conditions.

2.3.3. Uses of the Fitness Concepts

There are significant differences in what these fitness concepts represent, and in their time scales. Fitness and rate coefficients arise directly out of the process of natural selection, and therefore are always significant when natural selection proceeds; they are direct and fundamental measures of the speed of the process. On the other hand, natural selection does not necessarily lead to a change in trait value or frequency, or in increased adaptedness, adaptability, or durability; those are derived measures, or measures of a different process (evolution). For example, for a variety of fossil species, Van Valen (1973) showed that the probability of extinction is nearly independent of the age of the species. One might think that the probability of extinction in the next time unit would be smaller for older species than for younger species as natural selection improves the "fit" between species and their environments (Lewontin 1978). But this would *only* be true if the environment were absolutely constant; climate and geology change so markedly over millions of years that it might be sufficiently difficult to adapt to the changes so that extinction rates become essentially independent of age (Van Valen 1973; Lewontin 1978). Thus an important distinction among the concepts is that fitness and rate coefficients are properties of natural selection, but adaptedness, adaptability, and durability are properties of evolution (Table 1.1). Fitness, rate coefficients, and adaptedness apply to time scales of a generation or less, while adaptability and durability apply over many generations.

Fitness, rate coefficients, and adaptedness are predictable, given current conditions, while adaptability and durability depend upon the past and future history. Thus natural selection is predictable, whereas evolution is not, unless we were to have a complete understanding of geophysics and meteorology. This is one reason that some palaeontologists and systematists often downplay the importance of natural selection in evolution; randomly varying natural selection is not much different in effect from genetic drift. This needs further study by palaeontologists and palaeoclimatologists, as well as theoreticians.

2.4. TWO MORE USEFUL DISTINCTIONS

Sober (1984) makes the important distinction between selection *of* a trait and selection *for* a trait. (A third phrase is "selection *on* a trait"; this implies that natural selection as a "force," and should not be used.) If we speak of selection *for* a trait, this indicates that a direct cause-effect relationship has been identified for that trait. However, if we speak of selection *of* a trait, this is merely an empirical observation of the *effects* of natural selection (or some other evolutionary factor), or that a particular trait appears to be favored in natural selection. The distinction is important because there may not be a direct causal relationship between natural selection and the change in distribution of a given trait. For example, if two traits are genetically linked, if their loci interact, or if they are phenotypically correlated, selection for the first trait will result in a change in the distributions of both trait values. We can only say that there was selection *of* the second trait.

There is also an important distinction between demonstrating that natural selection has occurred or is occurring in a given population and demonstrating the validity of the definition of natural selection. Disproof of natural selection in any population does not affect its general validity, but the generality of the syllogism (Chapter 1) does not imply that natural selection always occurs. (For a discussion, see Sober [1984].) A related point is that just because natural selection is occurring now does not mean that it will continue to occur, or that it occurred in the past. It must be demonstrated at each time interval.

2.5. SUMMARY

When properly defined, natural selection is a syllogism rather than a tautology. Natural selection is a process which results from biological differences among individuals, and which may give rise to genetic change or evolution; it is not a "force" that "acts," and does not have an "intensity." By analogy with a chemical reaction process, natural selection has a rate and rate coefficients, which are estimated by fitnesses. The same analogy shows why "force" is an improper analogy; it is not a "force" that causes the change in reactant frequencies, but the chemical properties of the reactants. "Fitness" is a description rather than an explanation, and therefore it is not a tautology. At least five different meanings of "fitness" are found in the literature: fitness, rate of selection, adaptedness, adaptability, and durability. They differ significantly in what they represent; two apply to natural selection and three to evolution. Natural selection is predictable whereas evolution is not. Natural selection cannot explain the origin of new variants and adaptations, only their spread.

CHAPTER THREE

Methods for the Detection of Natural Selection in the Wild

> To expect that the intricacies of science will be pierced by a careless glance, or the eminences of fame ascended without labour, is to expect a particular privilege, a power denied to the rest of mankind; but to suppose that the maze is inscrutable to diligence, or the heights inaccessible to perseverance, is to submit tamely to the tyranny of fancy, and enchain the mind in voluntary shackles.
>
> Samuel Johnson, *The Rambler*, 9 July 1751

Many methods have been used to detect natural selection. Those that can be used in natural populations may be roughly divided into ten classes, and these are presented in Table 3.1. This chapter summarizes and compares these methods but does not describe them in detail; details may be found in the references cited in the discussion for each method. This chapter also discusses briefly the problems that are specific to each method; problems that are general to most or all methods will be discussed in Chapter 4. A brief discussion on how and why one should look for natural selection concludes the chapter.

The ten methods have different properties, and hence do not work equally well on any given species, nor do they yield the same kind of information. In fact, some are insufficient by themselves to demonstrate natural selection. This becomes clear upon examining their logical structure. Each method can be defined in terms of a null hypothesis and one or more alternative hypotheses, which may be accepted if the null hypothesis is rejected (see Table 3.1). With the exception of method IV,

Table 3.1. Summary of methods that have been used for detecting natural selection in natural populations

Null Hypotheses	Alternative Hypotheses
I. *Correlation with environmental factors*	
Unselected traits will vary independently of environmental factors.	Geographically varying selection results in a correlation between traits and their selective factors.
II. *Comparisons between closely related sympatric species*	
Independent variation among homologous traits of sympatric species.	Homologous traits of sympatric species are affected by the same kinds of selection: 1. Geographic correlations among homologous trait distributions if species do not interact. 2. Character displacement if species interact.
III. *Comparisons between unrelated species living in similar habitats*	
Independent variation among species living in equivalent habitats.	Analogous traits of species living in equivalent habitats are affected by the same kinds of selection and will converge: 1. In allopatry due to equivalent selection. 2. In sympatry due to interspecies interactions.
IV. *Deviation from formal null models*	
No selection.	Some form of selection.
Models:	
1. Hardy-Weinberg	Selection alters genotype frequencies.
2. Allele frequency distributions	Selection alters distribution and effective number of alleles.
3. Gametic phase disequilibrium or trait covariance	Selection causes nonrandom associations among: a. Alleles at different loci. b. Quantitative traits.

(*continued*)

TABLE 3.1 (*continued*)

Null Hypotheses	Alternative Hypotheses
4. Lewontin-Krakauer	Selection affects F_{ST} differently at different loci in the same population.
5. Genetic divergence	Selection affects variance of distance among populations.
6. Gene flow methods	
a. Decay of cline due to secondary contact with known age, gene flow, and starting frequencies.	a. Selection affects expected slope of cline.
b. Decay of secondary contact clines at several loci with known starting frequencies.	b. Selection will cause slopes to vary among loci more than that expected by chance.
c. Comparison of indirect gene flow measures at different loci.	c. Selection will cause heterogeneity in measures.
d. Comparison of sizes and mean values of differentiated areas.	d. Selection will cause areas to be larger or more differentiated than by chance.
V. *Long-term studies of trait frequency distributions*	
Traits vary at random in time.	1. Long-term stability, with less temporal variation than expected by chance. 2. Regular directional change.
VI. *Perturbation of natural populations*	
After perturbation the trait frequency distributions will change only by chance.	The trait distributions will diverge from what they were immediately after perturbation.

Perturbation may be by

1. Artificially induced changes in populations or their environments.
2. Temporary changes in the environment
3. Introduction into new equivalent and nonequivalent habitats.
4. Known abrupt changes in the environment.
5. Known gradual changes in the environment.
6. Known seasonal changes in the environment.

(*continued*)

TABLE 3.1 (*continued*)

Null Hypotheses	Alternative Hypotheses
VII. *Genetic demography or cohort analysis*	
Cohorts differing in trait frequency distributions differ only by chance in demographic parameters.	Significant relationship between trait value and demography; condition *b* holds.
VIII. *Comparisons among age classes or life-history stages*	
Age classes differ only by chance in their trait frequency distributions.	Natural selection results in differences in trait frequency distributions among age classes.
Methods: 1. Comparison of breeding and nonbreeding adults. 2. Comparison of juveniles and adults. 3. Comparison of a few age classes. 4. Comparison of live and dead individuals. 5. Fitness component analysis.	
IX. *Nonequilibrium prediction of condition* b *or its consequences*	
Null hypothesis is a particular selection model.	1. Inappropriate model. 2. No selection.
Methods: 1. Predictions from independently estimated fitnesses and known genetics. 2. Predictions about condition *b* for natural selection, derived from first principles of biology, biophysics, biochemistry, etc.	
X. *Equilibrium prediction of the outcome of natural selection*	
Null hypothesis is a particular selection model derived from optimization or fitness maximization models.	1. Inappropriate model 2. No selection.

these hypotheses are usually not explicitly stated in the literature. Most methods have no selection as the null hypothesis, but in methods IX and X selection is the null hypothesis. Some methods examine the outcome of natural selection while others specifically test for its conditions. Some can and others cannot distinguish between current selection in the study population and selection some time in the past. Thus, the methods vary in their directness and their ability to demonstrate natural selection in the wild.

The ten methods apply to both polymorphic and quantitative characters; the term "trait frequency distribution" refers both to the distribution of trait values for quantitative traits, and to the allele frequency array for polymorphic traits.

3.1. METHOD I: CORRELATION WITH ENVIRONMENTAL FACTORS

This is probably the commonest and oldest method. If natural selection occurs, then geographic variation in the selective factor will give rise to parallel geographic variation in traits. Null hypothesis: unselected traits will vary independently of environmental variables. Alternative hypothesis: geographically varying natural selection will result in a correlation between the selected traits and the selective environmental factors, after the effects of genotype-environment interaction have been removed. Thus a plot of a suspected environmental variable (X) against the trait of interest (Y), and a significant correlation coefficient between them, will suggest the presence of natural selection of that trait. Correlations may be found by simple (Ford 1975) or multivariate methods (Kahler et al. 1980; Piazza et al. 1981), and more complex geographic hypotheses can be tested by means of the Mantel test (Douglas and Endler 1983; Dietz 1983). Multivariate methods are particularly useful when more than one environmental factor or more than one trait (or both) is being studied.

Any area has environmental gradients going in various directions. If one were to throw a pencil on a map of the area it is almost certain that *some* environmental gradients will have components going parallel to the pencil; only perpendicular gradients will have none. If the pencil represents a transect of measured trait values through the species range, then it is very likely that *some* significant environmental correlations will appear by chance alone (Clarke 1975, 1978; Clarke et al. 1978). This is a serious problem if only a few localities have been sampled (Clarke 1975; Brown 1979; Clegg et al. 1978b), but it is not if there is a very close correlation over a large geographic area that follows geographically complex environmental variation. There are good examples in humans (Piazza et al. 1981) and fish (Endler 1978, 1982b).

It is important to use traits that are either independent of environmental effects or have known heritability, so that environmental effects can be statistically removed during the analysis. If there is a strong environmental component to the trait variation, then geographic variation in the environmental component inducing the phenotypic variation will be correlated with the phenotypic variation. In the worst case, where there is environmental induction of trait variation, a genetic component to the variation, but no natural selection of the trait, the correlation will be poor. To the naive worker, the correlation might be interpreted as imperfect because environmental noise is partially "hiding" the supposed correlation between geographic variation in genotypes and environment. It is very important to know the heritability of traits that are geographically correlated with environmental factors.

It is also important to understand the nature and form of any genotype-environment interaction. In a study of size and shape in *Cerion uva* snails, Gould (1984) found that the phenotypic covariance of many shell measurements (including different ontogenic stages) varied geographically. This results in different allometric relationships in different areas. Gould's scenario is that, in places within areas where conditions are better, the snails grow faster and larger, giving rise to differences in shape that are due to allometry rather than selection. The differences among areas in the traits' variance-covariance matrices impose a larger-scale geographic variation in shape on top of the local allometric effects. Therefore at least some of the overall correlation between environment and shape may result merely from environmental induction of size differences, which, through allometry, affect shape. The critical tests have not been done, but if the shape differences were indeed due to allometric effects of differential growth, it would be a good indication of how ignorance of genetics and development, in conjunction with method I, can give a very misleading picture of the dynamics of morphology and natural selection. Both the heritability and the nature and form of any genotype-environment interaction must be known for the proper use of method I, as well as for all methods.

A correlation may suggest a causal relationship, but this is not sufficient to demonstrate it. Unfortunately, it is all too common for only correlation evidence to be presented in the literature; this is very weak evidence for natural selection. If the correlation is not very high, sometimes modifications are made in the analysis. This may result in a series of progressive ad hoc hypotheses about the relationship between the trait and the environmental factor or factors (discussed in Lewontin 1974; Gould and Lewontin 1979; Clarke 1975; Clarke et al. 1978). On the other hand, if there is a strong correlation between complex geographic variation of trait and environment, or if the correlation is uniform over a very large area, then the evidence is somewhat stronger. An example is *Drosophila melanogaster* ADH and G3PDH allozyme frequencies, which have strong correlations with geographically complex patterns of rainfall (Oakeshott et al. 1982). But even if the correlation is very good, other methods are always needed to demonstrate a *casual* relationship.

Thus method I is indirect and can only suggest natural selection. It does not directly demonstrate the conditions for natural selection, and cannot show whether selection is currently present or happened some time in the past. Because it suggests possible causal relationships between traits and environmental factors, method I is valuable in suggesting further research on the environmentally correlated traits (Cain 1983; for examples, see Ford 1975 and Endler 1978).

3.2. METHOD II: COMPARISONS BETWEEN CLOSELY RELATED SYMPATRIC SPECIES

This method depends upon the assumption that relatedness of two species implies common response to the same environmental factors. It depends upon homology, so the species being compared must be closely enough related for it to be reasonable to assign homologies among their traits. Null hypothesis: independent variation is found among homologous traits of species living in the same area. Alternative hypothesis: homologous traits are affected by the same kinds of natural selection. There are two different alternative hypotheses, depending upon whether or not the species interact with each other.

3.2.1. Noninteracting Species

Alternative hypothesis 1: if the species do not interact, their homologous traits are expected to respond to the common environment in the same way, yielding similar and parallel geographic variation, after the effects of genotype-environment interaction have been removed. This will result in strong between-species correlations among suites of traits.[1] In studying single traits with simple genetics and function, it may be possible to compare more distantly related species—for example, geographic variation in melanism of moths, bugs, and beetles (Kettlewell 1973).

This method will only indicate selection if there is no introgression or gene flow between the species. If there is gene

[1] See Clarke 1975. Examples in Borowsky 1977; Carter 1967; Gill 1981; Hagen et al. 1980; Harrison 1977; Johnson 1974; Koehn and Mitton 1972; Lamotte 1951.

flow, then the two species will show parallel geographic variation due to the interchange of alleles rather than because they are responding to the same environmental variables (Clarke 1975). If two species are closely related, they are likely to have similar genetics and development—for example, similar linkage relationships, epistatis, and pleiotropy: their genetic variance-covariance matrices for the traits will be similar. If the two species were both split into two (or more) geographically concordant isolated populations by climatic changes, were differentiated at random while isolated, and came into secondary contact again, then the similarity in their variance-covariance structure may yield parallel geographic variation in the absence of selection. Common geographic history may produce similar trait variation and evolution; in fact, this is an assumption of vacariance biogeography (see discussion in Endler 1982a). Thus one has to be very careful in interpreting parallel trait variation in related species.

3.2.2. Interacting Species

Alternative hypothesis 2: if the species compete, their traits should diverge in the area of sympatry, reducing competition, while remaining more similar in allopatric populations. This is known as character displacement.[2] There are very few unequivocal examples (Bell 1976; Dunham et al. 1979; Schluter et al. 1985) because it is difficult to obtain critical data (Grant 1972, 1975; Levinton 1982). For example, Grant (1975) showed how an apparent case of character displacement can arise if two species clines (geographic gradients in trait distributions) that are roughly parallel (similar slope) but different in position, and there is incomplete sampling on one side of the sympatric zone. If only one or two samples are taken on one side of the zone and many on the other, then sampling will be enough to show the cline in one species, apparently no cline in the second

[2] See Brown and Wilson 1956; Grant 1972a, 1975; MacArthur 1972; Clarke 1975; Slatkin 1980.

species outside the zone of contact, and a displaced cline in the zone—false character displacement (Figure 3.1). This can be avoided by thoroughly sampling in, on both sides, and at some distance from the contact zone.

Other forms of between-species interactions, such as predation, mutualism, parasitism, or mimicry may also cause divergence. This does not seem to have been investigated in closely related species, probably because they tend to be in the same trophic level, and would bear further investigation. A possible example is the "race" between models and mimics, though, again, they tend to be in different families (Ford 1975). Harvey et al. (1983) suggest that competitive exclusion of species with particular intermediate phenotypes may yield apparent

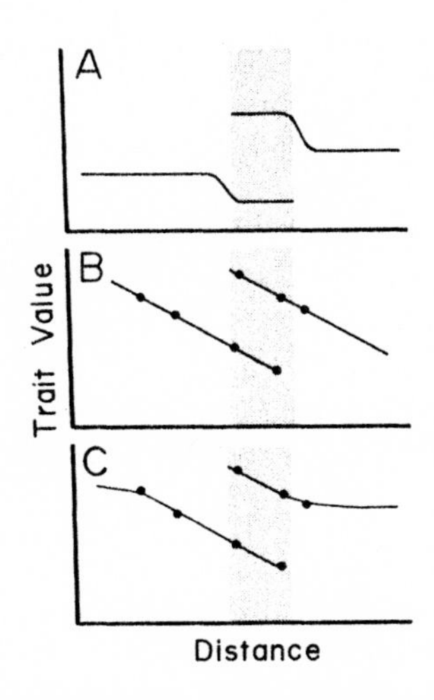

FIGURE 3.1. True and false character displacement. Each graph is a plot of trait value versus distance along a transect for two species; the shaded zone is the area where the two species are sympatric. *A*, in true character displacement the trait values of both species are displaced where they are sympatric. *B*, two parallel clines; sampling points indicated by dots. *C*, showing same points as in *B*; if these were the only places sampled, they might be mistakenly extrapolated and interpreted as character displacement, as shown by the lines.

character displacement, without the occurrence of natural selection in the remaining species. This might work with phenotypes within species but has not been investigated.

An acceptance of either alternative hypothesis is indirect evidence of natural selection because neither demonstrates conditions *b* (fitness differences) and *c* (inheritance) for natural selection, and because neither can distinguish between natural selection occurring during the study or at some time in the past. Unlike method I, this method will not suggest a causal mechanism for the observations; it can only suggest that certain traits are worthy of further study.

It is also possible to compare related species that live in different habitats or places (allopatric), and look for divergence or for a correlation between trait differences and habitat differences. In that case it is assumed that speciation was associated with invasion of different habitats. The habitat differences between the sister species are associated with differences in natural selection on homologous traits, leading to divergence. The major problem here is that genetic drift of the same traits also favors among-species divergence in allopatry; drift cannot be distinguished from natural selection. Therefore a comparison of related allopatric species is not informative.

3.3. METHOD III: COMPARISONS BETWEEN UNRELATED SPECIES LIVING IN SIMILAR HABITATS

This method is based upon convergent evolution in similar conditions, so the species must be sufficiently unrelated to exclude parallel evolution, or similar evolution due to common genetics, development, and history. Null hypothesis: the traits of unrelated species vary independently. Alternative hypothesis: analogous traits of unrelated species will converge because they are subject to the same kinds of selection. The alternative hypothesis takes two forms, depending upon whether the species are allopatric or sympatric.

3.3.1. Allopatric Species

Alternative hypothesis 1: analogous traits of species living in different places but in equivalent habitats will converge because thay are subject to the same kinds of selection (Darwin 1859). There are numerous and often quite remarkable examples for both plants and animals.[3] Extreme convergence of multiple suites of traits between species as distant as baleen whales and flamingos (mouth morphology and feeding behavior, Olson and Feduccia 1980), or fish and cephalopods (Packard 1972), is strong indirect evidence that the traits are not adaptively neutral. The convergence is often so extreme that it causes serious taxonomic difficulties, as in neotropical honeycreepers (Beecher 1951) and in tropical freshwater snails (Davis 1979).

This method has not been used very often, at least in part because it is difficult to define habitat equivalence. If one selective factor can be defined and measured, this method can work well. For example, geographically varying visual predation intensity affects the geographic variation in color patterns of two allopatric Poeciliid fish species in the same way, even though their predators are completely different (Endler 1982b).

3.3.2. Sympatric Species

Alternative hypothesis 2: analogous traits of unrelated species living in the same environment will converge because the species interact and are affected by a common selective factor. The best example of this is Batesian and Mullerian mimicry, where the convergence between models and mimics (or among models) can be remarkable for color patterns, odors, or, in the case of parasites, cell surface properties (Ford 1975).

[3] See Beecher 1951; Box 1981; Cody and Mooney 1978; Davis 1979; Emerson and Radinsky 1980; Emmons and Gentry 1983; Mayr 1963; Nevo 1979; Olson and Feduccia 1980; Olson and Hasegawa 1979; Packard 1972; Simpson 1944; Willey 1911.

In conjunction with experimental manipulation, mimicry can be one of the strongest lines of evidence for natural selection. Other forms of interspecific interaction can also cause convergence—for example, in traits used in interspecific territorial displays in fish (Hagen et al. 1980) and birds (Cody and Brown 1970). As mentioned in the discussion of method II (section 3.2.1), it is possible to compare unrelated sympatric species that do not interact, if we can be sure that the traits of interest have at least a *functional* (rather than a genetic) homology between the species; melanism is a good example (Kettlewell 1973).

As are methods I and II, method III is indirect, since it does not directly demonstrate the conditions for natural selection and cannot distinguish between natural selection occurring during the study or at some time in the past. Like method II, method III will not suggest a causal mechanism for the observations; it can only suggest that certain traits are worthy of further study.

3.4. METHOD IV: DEVIATION FROM FORMAL NULL MODELS

In this method a formal model is set up, assuming no selection, and the statistical consequences are predicted and compared with the data. Following Harvey et al. (1983) and Colwell and Winkler (1984), a null model eliminates the effects of some particular ecological or genetic process, in this case natural selection. There are six classes of null models, but not all apply to both quantitative and polymorphic traits.

3.4.1. Hardy-Weinberg

If the traits of interest are polymorphic, then the null model predicts genotype frequencies. The assumptions are (a) diploidy; (b) single locus with normal Mendelian genetics; (c) all genotype classes distinguishable; (d) sexual reproduction and random mating; (e) population size large enough for negligible genetic drift; (f) negligible mutation; (g) no natural selection;

(h) no gene flow; (i) no epistatic or linkage effects with other loci; (j) nonoverlapping generations. For n alleles at a single locus with frequencies $p_1, p_2, \ldots, p_n$, the expected genotype frequencies are given by the expansion of the multinomial $(p_1 + p_2 + \cdots + p_n)^2$. The distribution of deviations is discussed by Robertson and Hill (1984). Elaboration for several loci is possible if linkage disequilibrium is known (Ewens 1979; Elston and Forthofer 1977). If the loci are autosomal, another null hypothesis is no difference between the sexes. Although a rejection of the null hypothesis can mean selection has occurred, it can also mean a violation of any of the other assumptions; this method cannot by itself demonstrate selection. Additional problems include (a) it is a weak test statistically (Barndorff-Nielsen 1977; Brown 1970; Lewontin and Cockerham 1959; Schaap 1980; Ward and Sing 1970; Workman 1969); (b) deviations are hard to interpret unless zygote frequencies are known (Wallace 1958; Christiansen et al. 1973); (c) some kinds of selection lead to negligible deviations from expectation—for example, fecundity differences (Li 1959; Prout 1965–1971a,b); (d) the method may give rise to spurious frequency-dependence (Prout 1965–1971a,b) or heterozygote disadvantage (Wahlund effect, Wallace 1981), depending upon measurement conditions; (e) it ignores effects of sexual selection (Christiansen et al. 1973), though the effects of sexual selection can be allowed for in detailed studies (method VIII, Prout 1965–1971a,b; Christiansen and Frydenberg 1973; Ostergaard and Christiansen 1981). Analogous tests for quantitative characters are either not very informative or not practical; the number of loci contributing to the trait, and the magnitudes of their contributions, are usually unknown (Bulmer 1980; Ewens 1979).

3.4.2. Allele Frequency Distributions

If all alleles at a locus have been identified, and the allele frequencies are stationary, then we can predict the number of alleles expected under a balance between mutation to neutral alleles and loss due to random genetic drift (Ewens 1972, 1979; Waterson 1978). The expected number of alleles at a locus is positively related to $4N_e\mu$, where N_e is the effective population size and μ is the mutation rate to neutral alleles. At equilibrium, the quantity $4N_e\mu$ is estimated by the observed number of alleles k and the sample size r. This is used to predict F or the "homozygosity" for a given r. The effective number of alleles at a locus is $1/F$. The observed value of F is $\sum p_i^2$, where p_i is the frequency of allele i. The probability of a given observed value of F, given r and k, is most easily estimated (without laborious calculations) by means of the table in Appendix C of Ewens (1979). This is a more sensitive test than older methods. The null hypothesis yields a particular range of expected values for F, and it is a two-tailed test. If the observed F is greater than the upper value, it is likely that one or more of the alleles are disadvantageous and are being lost from the population, making somes alleles more common. If the observed F is smaller than the lower value, then some form of stabilizing selection (such as heterozygote advantage) is actively maintaining alleles in the population, making the frequency distribution of alleles more even. For further discussion of this and other methods, see Ewens (1979). Methods have also been devised for testing the distribution of alleles directly, rather than using the distribution of the effective number of alleles (Nei et al. 1976; Fuerst et al. 1977; Chakraborty et al. 1980; Kimura 1983).

3.4.3. Gametic Phase (Linkage) Disequilibrium and Trait Covariance

If two unlinked loci are in Hardy-Weinberg equilibrium, then the frequencies of coupling and repulsion gametes should be equal. Given enough time after a random deviation from gametic phase equilibrium, the two classes will be equal even if the loci are linked. If selection favors particular combinations of genotypes at different loci (linked or unlinked), then this should show up as gametic phase or linkage disequilibrium.[4] A similar argument can be made for correlational selection of quantitative traits (Arnold et al. 1986).

To obtain the most information on the nature of the disequilibrium, we may break it down into a number of different coefficients that estimate the contribution to disequilibrium by various combinations of alleles at two or more loci. These coefficients can be tested against the expected values (Weir 1979; Weir and Cockerham 1978, 1979). Unfortunately, very large sample sizes are required to achieve statistical significance even for moderate coefficients. Simpler methods exist (Brown et al. 1980; Malpica and Briscoe 1982), but are often less biologically informative and may be less sensitive (Brown et al. 1980). In an analogous fashion, geographic covariance among quantitative traits may also indicate selection (Bulmer 1980; Baker 1980; Johnston 1973; Hegmann and Dingle 1982); but as mentioned for method II, we can rarely eliminate the effects of genetic interactions that cause particular covariance patterns.

Gametic phase disequilibrium may not be a reliable indicator of selection since it can result from several forms of population structure such as population subdivision, breeding structure, secondary contact of two or more populations with few haplotypes, short periods of low N_e, and other aspects of random genetic drift.[5] Differences in recombination rates between *cis* and *trans* double heterozygotes ("repulsion" and "coupling") can also yield linkage disequilibrium (Clark and Feldman

[4] See Fisher 1930, 1939. Examples in Allard, Babbel, et al. 1972; Allard et al. 1972; Bodmer 1973; Brown et al. 1980; Cain and Sheppard 1954; Hamrick and Holden 1979; Levitan 1973b; Nevo 1978; Prakash and Lewontin 1971.

[5] Discussions and a variety of examples are found in Allard et al. 1972; Bodmer 1973; Bulmer 1980; Clark and Feldman 1981a,b; Clarke 1975; Ewens 1979; Feldman and Christiansen 1975; Hill and Robertson 1968; Nei and Li 1973; Prakash and Lewontin 1971; Sved 1968; Weir 1979; Weir and Cockerham 1978, 1979.

1981b). There are several other ways in which interactions among many loci, in conjunction with population structure, can result in apparent selection (Ewens 1979); these have not been demonstrated in natural populations, and demonstration would not be easy.

"Hitch-hiking" (Maynard Smith and Haigh 1974) is a situation in which a given allele changes in frequency as a result of linkage or gametic phase disequilibrium with another selected locus (summarized in Thompson 1977 and Ewens 1979). It can also give a false impression of selection at a particular locus (Clarke 1975)—for example, in barley *Hordeum* and oats *Avena* (Hedrick and Holden 1979). Similarly, if there is genotypic correlation among quantitative traits, then selection will appear to affect a trait directly, although it is actually only affected through its correlation with another selected character (a correlated response, Falconer 1981). If there is a phenotypic correlation among quantitative traits, this may also yield false or misleading evidence for phenotypic selection (Lande and Arnold 1983; Arnold et al. 1986); an example is shown in Figure 3.2. Effects of correlations among quantitative characters on phenotypic selection can be eliminated by means of multiple regression (Lande and Arnold 1983); this will be discussed further in Chapter 6. Both phenotypic and genotypic correlations can be independent and possibly even have opposite signs; both must be investigated for a complete study of natural selection.

3.4.4. Lewontin-Krakauer

The fixation index F_{ST} is an estimate of the probability that two alleles, drawn at random from a subpopulation, are identical by descent, and indicates the effect of present and past population structure (Wright 1931). It is the ratio of the observed variance of allele frequency among populations (s_p^2) to that expected from the mean allele frequency ($\bar{p}$) of all populations, or $F_{ST} = s_p^2/\{\bar{p}(1 - \bar{p})\}$. Many different loci should yield the same F_{ST} values because, being in the same organisms, they have experienced the same population history. Selection will tend to make some of the F_{ST} values deviate from the values expected from the null hypothesis (Lewontin and Krakauer 1973, 1975; Tsakas and Krimbas 1976; Nevo et al. 1975). A similar test has been devised for quantitative traits (Krimbas 1976). Unfortunately, there are serious problems in estimating the expected variance of F_{ST} and constructing a

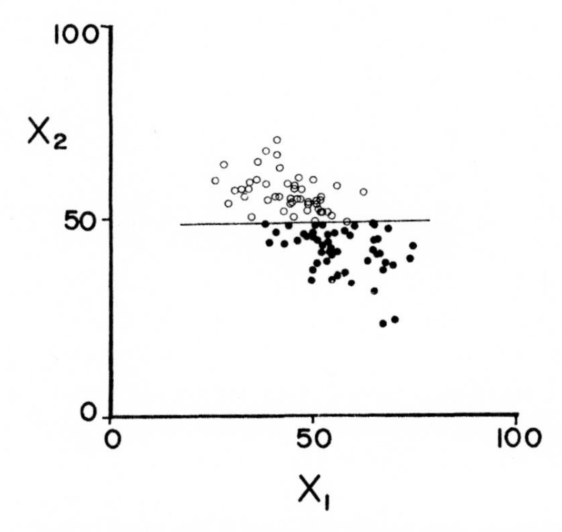

FIGURE 3.2. The effect of phenotypic selection on a correlated trait. One hundred individuals varying in two traits, 1 and 2, were drawn at random from a bivariate normal distribution with means of 50.0, standard deviations of 10.0, and a correlation of −0.7. The individual values for each pair of traits are shown as open or solid dots; the whole distribution may be regarded as the population before selection. Let only trait 2 be relevant to fitness (survival or mating), and for simplicity consider threshold (truncation) selection; the individuals above the horizontal line (open circles) have high fitness. Note how this results in a shift in the mean of trait 1 as well as trait 2, even though only trait 2 affects fitness directly; the mean of high fitness individuals increases for trait 2 and decreases for trait 1. This is distinct from and additional to the correlated *response* to selection, which results from genetic correlations among traits.

valid statistical test.[6] Worse, changes in population structure, such as fluctuations in effective population size, followed by reaggregation of formerly isolated demes or other forms of genetic admixture, can affect the distribution of F_{ST} values among loci (Robertson and Hill 1984). For quantitative characters there is the additional problem of correlation among characters, and the possibility of major differences between the phenotype and genotype variance-covariance matricies (example in Hegmann and Dingle 1982); this would yield quite misleading results.

3.4.5. *Genetic Divergence*

If isolated but formerly connected populations diverge at random, then we will expect a certain degree of genetic divergence among them at any given time since their separation. There are methods for both polymorphic and quantitative traits.

Genetic Distance. If many polymorphic loci are considered, a measure of the amount of differentiation between two populations is the genetic distance, a function of the differences among gene or genotype frequencies at each locus. There are a variety of methods of estimating genetic distance, with an equal variety of assumptions and properties (Bodmer and Cavalli-Sforza 1971; Nei 1973, 1978; Nei and Roychoudhury 1974; Reynolds et al. 1983). Good comparative summaries of the measures are found in Goodman (1974), Hedrick (1975, 1983), and Nei (1978). In the absence of selection (or other violations of Hardy-Weinberg), we expect isolated populations to diverge at all loci by genetic drift at an average rate, depending upon the effective population size. Thus, as time since separation increases, the average genetic distance among populations also increases. At any given time after a group of populations was part of one original population, the populations

[6] See Robertson 1975; Ewens 1979; Ewens and Feldman 1976; Nei and Chakravarti 1977; Nei et al. 1977; Nei and Chesser 1983.

will have an expected distribution of genetic distances, which is a function of time since separation, their effective population sizes, and (in some models) the mutation rate (Nei and Roychoudhury 1974; Reynolds et al. 1983). The various measures of genetic distance vary in their accuracy in estimating either time since divergence, or amount of differentiation since divergence. The inbreeding coefficient F (also called the coancestry coefficient θ, the probability that a random pair of alleles at the same locus are identical by descent) appears to be the best measure (for review and comments see Reynolds et al. 1983). If there is natural selection, this will affect the distribution of distance measures. Depending upon the spatial and temporal pattern of selection, and the number of loci affected by natural selection, the variance of genetic distance can be smaller, larger, or even equal to that expected by drift or by drift and mutation alone. This makes interpretation of an accepted null hypothesis difficult. A major difficulty in applying this method to natural populations is that we usually do not know the time since separation, or the average effective population size since separation. A second difficulty is that violations from Hardy-Weinberg other than selection will also affect the observed distribution of genetic distances.

Quantitative Divergence. Under genetic drift alone, for a given group of populations, the genetic variance of quantitative traits is expected to increase as a function of time, all other factors being held constant. Consider a species with observed phenotypic variation within populations v, realized heritability h^2, and effective population size N_e. If the observed phenotypic variation among n populations V is measured in a common environment, t generations after the separation, Lande (1977) shows that the expected distribution of $N_e V/h^2 vt$ is F with $(n - 1)$ and infinity degrees of freedom. A significant ratio indicates that natural selection has affected the rate of divergence since the separation time. Unfortunately this method can only be used to known times since separation, known generation times, and continuously known N_e.

3.4.6. *Gene Flow Methods*

There are four different methods: (1) If two formerly isolated populations came into secondary contact at a known date, if their starting differences in gene frequencies are known, and if the gene flow distance is known (root mean squared distance traveled by genes per generation, Endler 1977), then it is possible to predict the slope of the resulting cline, as it should decay at a constant rate. If there is selection, then the cline may be different from that expected by the null hypothesis (Endler 1977). (2) If there is secondary contact at an unknown date, or the gene flow distance is unknown, then an examination of many loci can indicate selection. In the absence of selection, the clines at each locus should be similar when corrected for initial differences between the populations. If there is selection at some or all loci, then the clines may be different in slope and in position (Bodmer 1973; Endler 1977). (3) Gene flow can be estimated indirectly at many loci in the same populations (for example, by the number of lethal alleles as a function of distance, Wallace 1981); if there is no selection, all loci should yield similar estimates. Heterogeneous estimates suggest selection at one or more loci (Adams and Ward 1973; Barker and East 1980). These methods allow estimation of fitness. (4) A species that is distributed over a sufficiently large area to allow isolation by distance will differentiate by genetic drift into a mosaic of patches of high and low gene frequencies or trait values. The size distribution of these patches is dependent on the gene flow distance and the neighborhood size N_e, as is the deviation (in patch value) of these patches from the mean of all patches (Endler 1977). A test for selection is to compare the observed size and value distributions of patches with those expected from a balance between gene flow and random genetic drift under the same population structure and density.

Geographically varying selection is likely to produce larger zones of strongly differentiated populations than drift and gene flow alone (Cain 1971; Endler 1977). To make this practical, more research needs to be conducted to determine the expected distribution of patches under various conditions.

In summary, method IV can be a good test for natural selection, so long as the null hypothesis is sufficiently well constructed so that its rejection indicates selection rather than some other factor. Although it does not test for natural selection conditions *b* (fitness differences) and *c* (inheritance) directly, it does test for deviations from current distributions expected if there were no selection; therefore this method can usually indicate if selection is currently occurring in the study area (Sections 3.4.1 and 3.4.6). In those cases some components of selection can be estimated.

3.5. METHOD V: LONG-TERM STUDIES OF TRAIT FREQUENCY DISTRIBUTIONS

This method requires data on trait frequency distributions over many generations, and the alternative hypotheses depend upon whether or not the traits of interest are at a stable equilibrium. Null hypothesis: the trait frequency distributions (mean and variance, or allele frequencies) should vary at random in time. Alternative hypotheses: (1) long-term stability (equilibrium), or (2) regular directional change (approach to equilibrium).

3.5.1. Long-Term Stability

Long-term stability of frequency distributions suggests a stable equilibrium, if the fluctuations are less than expected by genetic drift. This can arise from stabilizing selection (Bulmer 1980; or stabilizing selection with mutation, Lande 1975, 1976a), balanced polymorphism (Clarke 1975; Ford 1975), a balance between selection and meiotic drive, gene flow, or mutation (Ewens 1979; Lande 1976a), or simply a lack of genetic drift (Berry and Jacobson 1975; Cain 1971, 1983). Although so far it has been little used for this purpose, the fossil record may be useful here. There are clearly long periods of relatively little morphological change in some species ("stasis," Stanley 1979, 1982; but see Bookstein et al. 1978 and Gingerich 1983). Given the potentially great power of genetic drift for polymorphic (Ewens 1979) and quantitative traits (Lande 1975, 1976a) over such long time periods and the high genetic variability for these traits in living species, it seems unlikely that some form of selective equilibrium (such as stabilizing selection and mutation) does not account for the observed stability. But this is weak evidence for natural selection. Examples include bees (Alcock 1984) and sparrows (Zink 1983).

Long-term stability of a cline (geographic gradient in mean value or gene frequency) is stronger evidence for selection because, in the absence of selection, long-term stability of an entire geographic pattern is much less likely than stability at one or two localities (Endler 1977; Ellis et al. 1977b; Cain 1971, 1983). Examples include mice (Berry and Jacobson 1975); snails (Cain 1971, 1983) and several plant species (Ellis et al. 1977b; New 1978).

3.5.2. Long-Term Directional Trends

Long-term directional trends in trait frequency distributions suggest that the traits are not at a selective equilibrium, but that they are not drifting either. A long-term change can arise from directional selection, or an approach to an equilibrium that is shifting due to environmental changes. Good examples are grasses (Antonovics et al. 1971), great tits (Dhondt et al. 1979), *Drosophila* (Etges 1984), and several species of snails.[7] Cyclical long-term changes may also be informative, especially in the fossil record. For example, Kurten (1965) found that

[7] See Cain 1983; Clark et al. 1981; Clarke 1975; Clarke et al. 1978; Ford 1975; Ford and Sheppard 1969; Komai 1954; Murray and Clarke 1978; Wall et al. 1980.

some Pleistocene mammals shifted in size in parallel with several glacial and interglacial periods.

Unlike the first three methods, fitnesses (Clarke and Murray 1962; Dobzhansky 1970; Wall et al. 1980; Falconer 1981; Bulmer 1980) as well as epistatic effects (Clark et al. 1981) can be estimated from long-term regular trends, although these require explicit assumptions of the mode of selection.

An intriguing potential source of trouble is found in Feller (1968, chapter 3). In a discussion of coin tosses, he demonstrates that, if we assign the two sides of an unbiased coin the values of + 1 and − 1 and calculate the cumulative value (sum) of the results of many tosses, there is a surprisingly large probability that the value of the running sum will remain positive or negative for a long time. If we regard this as akin to genetic drift, where we are considering only the sign of the gene frequency change between generations, then this means that there is a nontrivial probability of getting a significant run of gene frequency changes in one direction. In computer simulations of genetic drift, it is common to obtain runs of positive or negative changes in gene frequency for many generations. To avoid this pitfall, it is necessary either to perform statistical runs tests on field data or, better still, to show directly that the conditions for natural selection are met by a detailed study of the ecology of the species (methods VII and VIII).

Method V (section 3.5.2) is more direct than methods I–IV because a rejection of the null hypothesis implies strongly that natural selection condition *b* (fitness differences) is met, although it does not demonstrate this condition directly.

3.6. METHOD VI: PERTURBATION OF NATURAL POPULATIONS

This method depends upon trait distributions responding to natural or artificial perturbations. It assumes that if natural selection occurs, then the trait frequency distributions are not at an equilibrium immediately after the perturbation. Null hypothesis: the population will not change after the perturbation, except through random genetic drift. Alternative hypothesis: the population will diverge from its new trait frequency distribution more than expected from drift, and in a consistent direction. The perturbation may be made either indirectly by a change in the environment, or directly by a change in the trait frequency distribution. If the environment is perturbed and this leads to a change in the trait distribution, then this suggests a cause-effect relationship. Implication of cause-effect relationships is not possible in the other methods except in method I.

Perturbations may be gradual or rapid. Rapid perturbations are more likely than gradual changes to yield measurable effects and cause-effect implications because they are more likely to overcome random effects. Perturbations may also be either temporary ("acute" or "pulse") or continuous ("chronic" or "press"). An excellent discussion of the different properties of temporary and continuous perturbation experiments in species communities is found in Bender et al. (1984). A most important point is that "pulse" experiments yield information only on direct interactions among species, while "press" experiments include both direct and indirect interactions. These arguments apply equally well to phenotypes within species. The six ways of studying perturbations summarized below are different combinations of these approaches.

3.6.1. Intentional Artificial Perturbations of Natural Populations

Intentional artificial perturbations of natural populations can provide very strong evidence for natural selection. This can be done either by perturbing the environment or the trait frequency distributions directly. The only difficulty is in producing a perturbation in the trait frequency distribution of known magnitude. A good example with known magnitude is Barker and East's (1980) *Drosophila* experiment; the initial and perturbed frequencies were known because it was possible to sample all breeding sites (cactus rots). This is one of the most direct and powerful methods of detecting natural selection, since, if after perturbation the trait frequency distribution continuously changes toward its previous value, it is difficult to explain the results in any way other than natural selection. If there are multiple equilibria in the system, it is possible for a perturbation to result in a regular change toward a new value, making interpretation more difficult.

3.6.2. Short-Term Natural or Human-Associated Environmental Perturbations

Short-term natural or human-associated environmental perturbations, such as storms, volcanic eruptions, epidemics, and logging can also be useful. Unfortunately, it is often difficult to know the magnitude of these perturbations, especially with respect to trait distributions, making the apparent response to such changes difficult to interpret. A possible example is the decreasing size of great tits (*Parus major*), which seems to be associated with an increase in the number of nesting sites, reducing the advantage of size in competition for nest sites (Dhondt et al. 1979). Presumably, given enough time with the same number of nest sites, the mean size would increase again. It would also be interesting to know whether or not mean size would increase if the nest boxes were removed. Another possible example is the increasing frequency of one color pattern allele (*typ*) in *Philaenus spumarius* spittlebugs, following a population crash of the vole *Microtus agrestis*, which affected the background color pattern and may also be a predator (Halkka et al. 1975). This example is noteworthy in that no change in background or *typ* frequency was observed in another island with no voles.

Most natural perturbations are not single events but occur continuously or seasonally, so it is difficult to distinguish between changes in distributions due to continuous perturbations, and changes that result from continuous approach to a shifting selective equilibrium. Except in special cases, such as well defined short-term natural events (as in Boag and Grant 1981), only experimental artificial perturbations (section 3.6.1) can yield unequivocal results.

3.6.3. *Natural or Human-Induced Introductions*

Natural or human-induced introductions into new habitats can also provide useful information: if there were no natural selection, the trait distribution of the introduced population should diverge from its ancestors only by chance. In this case, either the starting trait distributions must be known, or if not, the source population can be sampled again (Clegg and Allard 1972). Unfortunately, sampling from the source population will not yield data that will be able to distinguish selection from the founder effect (Mayr 1963), so the starting trait distribution really must be known. If the source population is not well known, but the introduced populations have diverged geographically from the source in various directions, then this is also good evidence of selection, provided there was a single source area. Good examples include natural or artificial introductions of *Drosophila melanogaster* (Oakeshott et al. 1982), guppies *Poecilia reticulata* (Endler 1980), house sparrows *Passer domesticus* (Baker 1980), oats *Avena* (Clegg and Allard 1972; Jain and Rai 1980), several species of grasses (Snaydon 1970), and pathogens with subsequent changes in both host and pathogen (Fenner and Ratcliffe 1965).

3.6.4. *Known Abrupt Changes in the Environment*

The introduction of pesticides and the subsequent and repeated evolution of pesticide resistance by the "pests" provide very strong evidence for natural selection, and usually make it possible to assign a direct causal relationship between the factor and the response. Examples include resistance in a wide variety of pathogens, fungi, plants, invertebrates, and vertebrates to a wide variety of pesticides.[8] The rapid response to the appearance of each new pesticide implies very strong selection, as does the evolution of disease or pathogen resistance after epidemics and outbreaks (Flor 1956; Fenner and Ratcliffe 1965; Day 1974). In host-parasite evolution, the rapid response is often in both host resistance and parasite virulence, as in the *Myxoma* virus and rabbit system (Fenner and Ratcliffe 1965).

3.6.5. *Known Gradual Changes in the Environment*

Known gradual changes in the environment may also be used to detect selection. A classic example is the increase in the frequency of melanic (dark) forms of many species of insects following the increase in air pollution in Britain, which was followed by a reduction in melanic frequency, as clean-air laws were passed (Kettlewell 1973; Creed 1971). Another fruitful line of research is on the effects of changing patterns of land use (Cameron et al. 1980a,b). A problem with more gradual environmental changes, is that it may be more difficult to identify the precise factors causing the change, and experimental methods will have to be used to verify such hypotheses. As species go extinct, it is also possible that their competitors will show character release as they invade the vacated niches (Grant 1972a), and their predators and parasites may also show shifts. These phenomena are best studied experimentally (Harvey et al. 1983).

3.6.6. *Known Seasonal or Periodic Changes in the Environment*

Known seasonal or periodic changes in the environment that are tightly correlated with changes in trait distributions

[8] See Brown 1978; Brown and Pal 1971; Watson and Brown 1977; Wharton and Norris 1980; Greaves and Ayers 1967, 1968; Bishop et al. 1977; Partridge 1979; Georghiou 1972; Holliday and Putwain 1980; May and Anderson 1983.

may also indicate selection.[9] However, because so many environmental factors change seasonally it will be very difficult to identify the factors directly causing the changes. Periodic epidemics or pest outbreaks may provide more specific information, as they are usually associated with specific genetic resistance to the particular strain of pathogen or pest that caused the outbreak. This yields a strong correlation between the genotypes of both hosts and parasites in both time and space.[10]

In any perturbation study, false apparent selection may occur if the trait frequency distribution is unknown before and immediately after the perturbation or if there are improper controls for introductions. If by chance the distribution after an introduction or perturbation is different than it is thought to be and there is no selection, then there will be an apparent change in distribution that would be falsely interpreted as the result of selection. This can be avoided by careful sampling immediately after an introduction or perturbation, but would be a serious problem in the study of a natural or man-associated introduction with unknown founders. It is mitigated in studies of introduced populations where extensive divergence in all directions from the unknown ancestors has taken place. An example is the house sparrow *Passer domesticus*, which has spread and markedly differentiated throughout North America (Johnston and Selander 1971) and New Zealand (Baker 1980).

Clearly, perturbations provide direct evidence for natural selection, though they do not formally provide direct evidence for conditions *b* and *c* (fitness differences and inheritance). However, if used in conjunction with cohort analysis (method VII) or age-class comparison (method VIII), method VI can provide

[9] Examples and discussions in Bishop 1969; Dobzhansky 1941, 1970; Fontdevila et al. 1983; Gallagher 1981; Heath 1974; Levitan 1973a; Merrell and Rodell 1968; Samallow 1980; Tonzetich and Ward 1973; Varvio-Aho et al. 1979.

[10] See Anderson and May 1981; Flor 1956; May and Anderson 1983; Clarke 1975; Alstad and Edmunds 1983a,b; Edmunds and Alstad 1978, 1981.

the strongest and most direct evidence for natural selection. Perturbation studies may be more likely than methods VII and VIII to detect natural selection if the population of interest is at or near equilibrium; this may be true for many species.

3.7. METHOD VII: GENETIC DEMOGRAPHY OR COHORT ANALYSIS

This can be the most laborious method because an attempt is made to obtain detailed information on the complete demography of the population with respect to each trait value or phenotype class. This method is sometimes known as "horizontal" or "longitudinal" study, but these terms are confusing or ambiguous, hence the plain English title used here. By gathering detailed data on individuals, data can be obtained on survivorship, fertility, fecundity, mating ability, and so on. Data on parents and offspring can also provide information on genetics (condition *c* for natural selection, inheritance). Data are best gathered from individually marked individuals, though some information can be gained by giving all members of the same cohort the same mark.

Null hypothesis: cohorts differing in trait frequency distributions differ in demographic parameters only by chance. Alternative hypothesis: particular demographic patterns are associated with particular trait values (condition *b*, fitness differences); the demographic parameters of particular cohorts depend upon their trait distributions. This is the most direct method for detecting natural selection because it specifically tests for condition *b*, and can also yield data on *c*. For this reason, it can also provide direct estimates of the rate of selection. Some methods (Lininger et al. 1979; Farewell and Dahlberg 1983) can be used merely to demonstrate selection, while others can be used to measure selection coefficients or differentials (see Chapter 6).

Method VII requires species in which it is practical to estimate the complete demography of a large fraction of the population. Most commonly it is practical to obtain data only on survival by marking and repeatedly sampling individuals, and methods are rapidly improving (for examples, see Bart and Robson 1982; Southwood 1978; Manley 1974). With more care, other components of condition *b* can be investigated. For example, fecundity may at first seem simple to sample (clutch size); but without a complete record of all or most individuals, one cannot know how many clutches are laid (or sired) per individual. Examples of cohort analyses include butterflies (Hoffman and Watt 1974; Sheppard 1951b), moths (Kettlewell 1973), land snails (Sheppard 1951a; Wolda 1963), frogs (Howard 1979; Arnold 1983a; Arnold and Wade 1984b), and humans (Bodmer 1968, 1973). This is the best way to treat data for intercorrelated quantitative characters (Lande and Arnold 1983; Arnold and Wade 1984a; Arnold et al. 1986).

This is the only method in which it is possible to obtain complete lifetime fitness estimates, which are required in order to predict changes in or equilibria of trait frequency distributions. Yet, because of the time and work involved, very few attempts at lifetime fitness measurement have been made. Two examples are tits *Parus major* (McGregor et al. 1981) and red deer *Cervus elaphus* (Clutton-Brock et al. 1982), although both are concerned mostly with nonheritable variation. Other examples are summarized in Clutton-Brock (1983).

3.8. METHOD VIII: COMPARISONS AMONG AGE CLASSES OR LIFE-HISTORY STAGES

This method is used when it is impractical to do cohort analysis. Instead of keeping track of individuals, simultaneous samples are taken of all age classes or life-history stages. For this reason, method VIII is sometimes known as the "vertical" or "cross-sectional" method, but these names are confusing or ambiguous, hence the plain English title used here. Samples of many and preferably all age classes or stages must be taken at the same time and place, preferably with repeated sampling from many localities. A comparison is made among the trait frequency distributions of different age classes or life-history stages.

Null hypothesis: age classes (or life history stages) differ only by chance in their trait frequency distributions. Alternative hypothesis: natural selection results in significant differences among some age classes; selection is greatest between the age classes with the greatest differences. As in method VII, this is a direct method of demonstrating ongoing natural selection; it specifically tests for the presence of condition *b*. As a result, it is possible to obtain estimates of fitnesses. If the study includes data from parents and their offspring, then information is also obtainable for condition *c*.

This method grades into the previous method (VII, Cohort Analysis), especially if sampling of the age classes is done repeatedly, or if all members of the same cohort are given the same mark. The major difference between the two is that in age-class comparison there is no data on the relative success of individuals, while there can be such data in cohort analysis if individuals, rather than cohorts, are individually marked and studied for the rest of their lives. Age-class comparisons infer condition *b* for natural selection by differences in trait frequency distributions among age classes, while cohort analysis measures condition *b* directly. There are approximately five variants of method VIII.

Comparisons may be made between breeding and nonbreeding adults in annual species (as in Mason 1964), between juveniles and adults (Weldon 1901; Berry et al. 1979; Dowdeswell 1961), or other life-history stages such as predispersal and postdispersal individuals. Comparisons can also be made among two or more age classes (as in Berry and Crothers 1968; Fujino and Kang 1968; Hiorns and Harrison 1970). There are similar methods for quantitative characters.[11] Comparisons can be made between living and dead individuals taken from the same place at the same time (Bumpus 1899; Sheppard 1951a; Richardson 1974). In a sense the latter is cohort analysis, method vii, but with no additional data on the live individuals.

One of the most efficient methods of detecting natural selection on polymorphic traits is fitness component analysis.[12] This is done by sampling as many age classes and life-history stages as possible, and doing it several times within a generation. It is still better to sample the same locality over several generations, and also, if possible, study more than one locality. By this intensive sampling, the effects of natural selection can be decomposed into components due to sexual, gametic, fecundity, and viability selection, and other components that are specific to particular life-history stages. These components can be divided into further elements: (1) sexual: mating ability, mate preference; (2) gametic: gamete competition, nonrandom union of gametes, meiotic drive; (3) fecundity: gamete production rate, zygote production rate; (4) viability: survival as zygotes, other life stages, developmental rate, age of reproduction, etc. As in method vii, data on parents and offspring also provide data on condition *c* (inheritance). Good examples are found in fish *Zoarces* (Christiansen 1977, 1980; Christiansen et al. 1973, 1977), and barley *Hordeum* (Clegg et al. 1978a). A useful modification for field selection experiments, and methods for esti-

[11] For discussions and examples, see Arnold 1983a, Arnold and Wade 1984a,b; Bodmer 1973; Beatson 1976; Haldane 1954; Leamy 1978; Lowther 1977; Manley 1975; O'Donald 1971, 1973; Van Valen 1965a, 1967; Van Valen and Mellin 1967.

[12] See Christiansen and Frydenberg 1973; Clark and Feldman 1981a,b; Clark et al. 1981; Nadeau and Baccus 1981; Prout 1965–1971a,b; Ostergaard and Christiansen 1981; summary in Hedrick 1983.

mating variances of selection coefficients, are found in Manley (1974).

Fitness component analysis was devised for polymorphic traits, but additional methods have been developed for quantitative traits. Total fitness can be decomposed into components for various life stages by an analysis of covariance and multiple regression of fitness on trait values (Arnold 1983a,b; Lande and Arnold 1983; Arnold and Wade 1984a,b; Arnold et al. 1986). Because quantitative characters are frequently correlated with each other, Lande and Arnold (1983) and Manley (1976, 1977) developed a method of decomposing the effects of selection on a particular character into components due to selection directly affecting that character and another component due to correlated effects of selection affecting other characters (see Chapter 6). Their method works better for cohort analysis (method VII) because in that case fewer assumptions are needed; for example, when studying size-related traits, it is unnecessary to assume a particular amount of average growth between age classes when it can be measured directly in cohorts.

In fitness component analysis, the segregation of Mendelian traits allows very efficient estimation of components of selection besides viability, especially mating ability and fecundity. Even meiotic drive and epistatic effects may be estimated (Clark and Feldman 1981a). Mendelian traits are also especially useful for difficult field situations in which it may not be easy to capture all life stages, or for that matter, observe mating or egg laying. For example, information on mate choice, fecundity, survivorship, and genetics can be gained when it is possible to associate one or two parents with their young (Christiansen 1980; Christiansen and Frydenberg 1976; Christiansen et al. 1973; Ostergaard and Christiansen 1981). The identification of parents and hence mating frequencies by means of multiple polymorphic loci can then be used to investigate many components of selection on quantitative traits in the same individuals. For example, if ten polymorphic loci are used to identify the fathers in combination with known combinations of mothers and offspring, then this can be used to define the mating structure. McCracken and Bradbury (1977) provide a good example of parental assignment and the estimation of mating frequencies using polymorphic traits. In addition, one or more quantitative trait distributions can be compared between the mated and unmated individuals of both sexes. There are many other possibilities.

3.9 METHODS IX AND X: PREDICTIONS ABOUT NATURAL SELECTION

These methods attempt to predict condition *b* for natural selection, changes in trait frequency distributions, or equilibrium distributions, on the basis of known properties of the traits, the environment, or both. Methods IX and X are logically very different from the others because the null hypothesis is a particular selective model rather than one with no selection. As a result, two different alternative hypotheses are always indicated if the null hypothesis is rejected: (1) the selection model is incorrect or inappropriate, or (2) there is no selection. Therefore, using only this method, one cannot prove that natural selection has occurred or is occurring; it becomes a matter of opinion: "neutralists" would favor "no selection," while "pan-selectionists" would favor "wrong model" (Lewontin 1974).

There is a serious potential problem in using this method. If one were to modify progressively the null hypothesis each time it is rejected, on the assumption that the trait of interest was influenced by natural selection, there is a real danger that the process will break down into an exercise of ingenuity in devising ad hoc hypotheses (Lewontin 1972, 1974; Gould and Lewontin 1979; Brady 1979). This is a major reason for the accusation that natural selection is "not testable,"[13] but it only applies to an exclusive and uncritical use of methods IX and X.

A good example of the proper modification of hypotheses is found in Charnov and Skinner's (1984) discussion of clutch size. Lack's original theory (Lack 1947a,b; Charnov and Krebs 1974) suggests an optimum clutch size because, if the clutch is smaller, fewer offspring will be produced; but if it is larger, the declining fledgling success with increasing clutch size will also result in fewer breeding offspring. Clutch sizes of birds (Charnov and Krebs 1974) and parasitoid wasps (Charnov and Skinner 1984) are usually found to be less than that predicted by Lack's model. The poor fit can be improved, not by an arbitrary ad hoc modification of the hypothesis, but by considering new data and more of the biology of the species. Lack's model did not account for three factors: (1) a negative correlation between mother's survival and either her clutch size or egg-laying rate; (2) an incomplete measure of offspring fitness; (3) the fact that offspring production per host is an incomplete measure of maternal fitness (Charnov and Skinner 1984). It is perfectly reasonable to make the model more realistic by including more known factors. For further discussion see Williams (1973), Wassermann (1978), and Oster and Wilson (1978).

If enough is known about the biology of a species, it is possible, at least in principle, to make detailed predictions about the conditions or outcome of natural selection on certain traits, either over one or a few generations, given the present trait frequency distribution (method IX) or at equilibrium (method X). These models attempt to deduce general properties of bio-

[13] See Lewontin 1972, 1974; Gould and Lewontin 1979; Gould 1982; Popper 1963–1974; Peters 1976; Eldredge and Cracraft 1980; Platnick 1977; Rosen 1978; Rosen and Buth 1980; Cox 1981; Pearson 1981; Macbeth 1971; Bethell 1976; Brady 1979; Tuomi 1981.

logical systems, based upon generalized assumptions about the direction and approximate rate of natural selection. They attempt to describe *why* organisms have or can evolve in certain ways, not *how* they have done so.

3.10. METHOD IX: NONEQUILIBRIUM PREDICTIONS OF CHANGES IN TRAIT DISTRIBUTIONS

This method works directly on condition *b* for natural selection to predict short-term changes in trait frequency distributions. There are two approaches. The first depends upon independently derived specific data on fitnesses, and the second is based upon general models that utilize first principles of biology to predict the parameters of condition *b*.

3.10.1. Method IXa: Predictions from Independently Estimated Fitness

If we have good estimates of selection coefficients (Wallace 1981) or differentials (Falconer 1981), we can predict short-term changes as well as the equilibrium (if any), even if we do not know *why* there is selection (Lewontin 1974). Although this has been most successful in laboratory populations (Falconer 1981; Dobzhansky and Pavlovsky 1953; Anderson 1964; Lewontin 1962; Lewontin and Dunn 1960; Lyttle 1979), it has also had some success in the field using laboratory estimates (Bell and Haglund 1978; Heath 1974; Lacy 1978). If precise estimates are impractical, more general predictions are possible about approximate magnitudes. General predictions have been especially successful in the analysis of host-parasite systems (Fenner and Ratcliffe 1965; Anderson and May 1981; May and Anderson 1983), and for the balance between selection and gene flow (Endler 1977). The effects of wind direction, pollen flow, and heavy metal tolerance in plants are examples (Antonovics et al. 1971; Karataglis 1980a).

3.10.2. Method IXb: Nonequilibrium Predictions about Condition b *from First Principles*

This method uses known properties of anatomy, physiology, energetics, behavior, etc., to set up predictions about the details of condition *b* for natural selection. These can then be used to predict short-term changes in trait frequency distributions, given their current distributions. It can also be used to predict equilibrium distributions, assuming constant conditions, but this is not necessary. This method does not assume that optimization or fitness maximization is necessary or even possible, nor does it assume that the population is at or near equilibrium, it simply asks whether a particular trait value is functionally better (or performs better, Arnold 1983b) than another.

This method may also ask whether or not the trait is the optimum from the point of pure functional design (without biological or historical constraints); but what is important in this method is what is *relatively* better design. The assumption here is that a trait value that is functionally superior to others is associated with disproportionate representation in the next generation, or, if there is some counteracting factor, only a shift in its distribution among age classes. In the few cases where this assumption has been tested, it appears to be valid (Endler 1980; Antonovics et al. 1971). Note that the object is not merely to state that a particular phenotype is "designed well" in the sense of Gould (1980, 1982), but functionally "designed" better than another for a particular purpose (Charnov 1982). As mentioned in the willow example (Chapter 2), the predictions must include as many components of fitness as possible, or they are unlikely to be successful.

Unlike method IX*a* (section 3.10.1), method IX*b* can get at the reasons for natural selection, since it starts from the biological details that can lead to condition *b* for natural selection. Thus, in addition to dealing with the causes of natural selection (rather than the effects), this method deals with "tactics" rather than "strategies".

Color patterns provide a good example of this method. If a color pattern serves only to prevent or delay detection by visually hunting predators, then we can predict its properties from (1) the biochemical ability of the prey to make certain colors; (2) the lighting conditions—intensity, spectral distribution, and air or water clarity when the prey is most vulnerable to predation; (3) the predator's visual abilities; and (4) background color patterns. This yields very specific predictions about the matching of animal and background color patterns at particular times and places, and these are easily testable (Endler 1978). If the color patterns serve other functions, such as thermoregulation or intraspecific communication, then the predictions can be modified by reference to geometry, optics, and visual physiology (Endler 1978, 1980, 1983, 1984). A direct measure of crypsis (Endler 1984) can indicate whether one color pattern is better than another, and also how close to the ideal it is.

There is a sizable literature on functional design (for example, biomechanics, Gans 1974; and especially swimming in fish, Webb 1975), yet students of natural selection have made little use of this literature. A notable exception is Riddell and Leggett's (1981; Riddell et al. 1981) study on *Salmo salar*. They worked out the heritabilities of various body-shape traits, and utilized information on stream flow and the hydrodynamics of swimming to predict geographic variation in body shape. Two more examples are Gross's (1978) study of the structure and function of spines and lateral plates in predator defense by sticklebacks, and Arnold's (1983b) discussion of snakes feeding on large prey items. (Arnold also provides a useful summary of an approach to method IX*b*.) Significant progress is being made in predicting optimum leaf morphology and branching patterns in plants (for example, Givnish 1979), but this has not yet been combined with demonstrations that variants that are predicted to be favored are actually those favored by natural selection. Some progress has also been made in defining physiological and kinetic differences among biochemical variants, but with very few exceptions (sickle cell in man—Allison 1964, 1975; Ingram 1957; G6PD in man—Allison and Clyde 1961; Luzzatto et al. 1969; PGI in *Colias* butterflies—Watt 1983; Watt et al. 1983; LAP in *Mytilus* mussels—Hilbish and Koehn 1985), these differences have not been related to mechanisms or direct demonstrations of natural selection (for summaries, see Koehn et al. 1983; McDonald 1983; Zera et al. 1985). For the most part the functional design literature is concerned more with the "how" than the "why" of design (for a perceptive insight and discussion, see Mayr 1961). This will change as we obtain more data on genetic variation for these traits and their relationships to the organisms' ecology.

3.11. METHOD X: EQUILIBRIUM PREDICTIONS ABOUT TRAIT FREQUENCY DISTRIBUTIONS

This method works on the final outcome of natural selection, assuming that all environmental conditions remain constant and that there is enough genetic variation to reach an equilibrium. Often it also makes some assumptions about optimization and fitness maximization. Predictions for equilibrium distributions are made from models that assume that the number of breeding offspring (or some other measure of fitness) is maximized in natural selection (Williams 1966; Maynard Smith 1978; Charnov 1982). Well-formed models consist of four parts: (1) a description of the possible distributions of trait values and other variables; (2) a set of alternative strategies; (3) a set of fitness criteria; and (4) a set of constraints (after Oster and Wilson 1978). Known biological relationships between the trait value and the number of breeding offspring are examined, and the trait value that gives the maximum number of breeding offspring is used for the predicted value. Often there will be the additional assumption of limited energy or resources, or of counteracting effects of interacting traits. In that case the optimum combination is calculated on the basis of maximizing the number of breeding offspring within the stated constraints. The stability of the predicted equilibrium is then estimated. It is quite possible for the system to be chaotic or unstable, especially when constraints are included; this makes prediction more difficult (Oster and Wilson 1978).

This method has been criticized (for example, Lewontin 1979) because of its assumption that something is being optimized or maximized (or minimized), and that there is a direct and significant correlation between what is being maximized and fitness. However, there is actually some direct evidence that traits may be optimized to maximize fitness (Perrins 1965; McGregor et al. 1981; Ware 1982). A related problem is the falsity of the assumption that fitness should always be maximized (Wright 1948; Sacks 1967), especially when considering more than one trait at a time (Moran 1964; Wright 1948, 1967), as is typical of optimization arguments. Although the genetic models that show the nonuniversality of fitness maximization are often very specialized and unrealistic, we really do not know enough about the actual interactions among loci and traits to discount their results for real organisms. For example, at least one case is known for yeast (*Saccharomyces cerevisiae*) in which fitness clearly decreases as a result of natural selection (Paquin and Adams 1983). On the other hand, it is unfortunate that most optimization models are frequency-dependent (Charnov 1982), because in frequency-dependent models fitness is not always at a maximum at equilibrium. It is also frequently forgotten that optimization models can only be used to find the optimum for the particular set of conditions of the population studied and cannot be used for long-term predictions about natural selection or evolution; environments change in evolutionary time. For an unusually balanced and detailed critique of optimization models, see Oster and Wilson (1978, Chapter 9).

Many of the better optimization studies start from an *observed* relationship between trait variation and some component of fitness in the field, rather than from an arbitrary relationship. In this case, natural selection is demonstrated and then used to predict the equilibrium trait distribution. In using this process, one sometimes forgets that the similarity between the predicted and observed distributions is not really a test for natural selection, since the relationship between the trait and fitness has already been demonstrated (condition *b*!). If the relationship comes from the laboratory, or another species, then this circularity is mitigated.

This method has had notable success when the traits of interest have a very large effect on fitness. For example, there is a remarkable agreement in detail between observations and predictions from optimization models for sex ratios and sex change in a great variety of animals and plants (Charnov 1982) and for clutch size (Charnov and Skinner 1984). Predictions have also been successful for the relationships between energetics, swimming speeds, and growth in fish (Ware 1975, 1978, 1982). Optimization models have also been applied to behavioral traits (the Evolutionarily Stable Strategy, Maynard Smith 1978), such as mating strategy in dung flies (reviewed in Parker 1983) and other animals (see Maynard Smith 1978, and papers in Bateson 1983 and Krebs and Davies 1978), where it can be a very useful conceptual framework. As Oster and Wilson (1978) suggest, "optimization models are a method for organizing empirical evidence, making educated guesses as to how evolution might have proceeded, and suggesting avenues for further empirical research."

3.12. HOW TO DETECT NATURAL SELECTION IN THE WILD

The ten methods vary in their ability to demonstrate directly natural selection; some test directly for conditions *b* and *c* while others test the predicted outcome of selection. Many are not sufficient by themselves to detect natural selection, and therefore a combination of methods is the best approach. An efficient way to proceed is to: (1) Find traits that may be affected by natural selection and environmental factors that may affect them; (2) demonstrate that natural selection affects those traits; (3) demonstrate causal factors that result in natural selection. If nothing is known about the biology of a species, then method I is a good starting point, as it indicates which traits may be affected by which environmental factors, but it does not demonstrate selection. Methods VI–VIII can be used to demonstrate natural selection directly, but they do not indicate why selection occurs. An experimental perturbation (method VI) of the environment rather than trait distributions may indicate a causal relationship in addition to providing a direct demonstration of natural selection. Once methods I and VI–VIII have been utilized, and enough is known about the species' biology and ecology, then method IX or X can be used to explore the causes of natural selection, which arise from biological and ecological differences among trait values. A demonstration of natural selection is not really complete unless the biological reasons for selection are discovered.

There are some people (e.g., Gould and Lewontin 1979) who suggest that it is very poor practice to assume that any given trait is adaptive or subject to natural selection. This criticism is directed at the uncritical use of only two methods of detecting natural selection (methods IX and especially X). Aside from the fact that proper studies do not use only these methods, there is a confusion here between *why* and *how* a study is done. A study is done to ask and answer a question, such as: Is a given trait subject to natural selection? This places no constraints on how the question is asked or how the hypothesis is constructed. Only two (IX and X) out of the ten methods ask the question by initially assuming the answer is "yes"; the other eight assume it is "no". Those with a slavish devotion to, and an incomplete understanding of, the "hypothetico-deductive" method of scientific investigation, have apparently forgotten where the hypothesis to be tested comes from in the first place; they have confused the generation of the hypothesis with the method of logical inference in testing it. By their insistence on never using methods IX and X and only using methods IV–VIII they are essentially making the nonadaptive hypothesis untestable and unprovable (a null hypothesis can never be "proven"). Ironically, this is the same fault they say we should avoid in the adaptive hypothesis. It is essential to use both forms of null hypothesis to explore the functions of traits.

If we did not use natural selection, adaptation, or functional utility as an initial working hypothesis, we would never learn anything about the biology of organisms (Cain 1964). If we assume that characters are nonadaptive and artifacts of history, constraints, or pleiotropy, then no investigations will be made, except for conditions *a* and *c*, to the impoverishment of biology. The assumption of nonadaptation can easily give rise to the construction of very plausible nonadaptive fairy tales (as in Gould 1984), encouraged by the same faults discussed in Gould and Lewontin (1979) for excessive devotion to adaptation. None of the studies summarized in Chapter 5 would have been done if the traits were assumed to be selectively neutral. It is certainly correct that the usefulness of a trait must not be taken for granted, but equally, its *uselessness* cannot be taken for granted either! Cain (1964) gives several examples of apparently trivial traits that have significant adaptive value upon close examination, but which could easily not have been investigated at all. A most interesting example is the Y-shaped bar of chitin on *Polyxenus* millipedes, which allows increased maneuverability without increasing leg length; increased leg length and low maneuverability are disadvantageous in tight places (Manton 1956, 1959; Cain 1964). The assumption that traits are subject to natural selection (but not necessarily adaptive) is much more productive than the assumption that traits are selectively neutral. This does not ignore the possible importance of genetic, developmental, or phylogenetic constraints—these should be explicitly placed in the models used in methods IX and X. The assumption that natural selection affects a trait has nothing to do with the logical structure of testing the hypothesis.

3.13. SUMMARY

Ten kinds of methods have been used to detect natural selection in the wild, but not all are capable by themselves of demonstrating its existence. The ten methods have different properties and hence do not work equally well on any given species, nor do they yield the same kind of information. Methods I–VIII have no selection as the null hypothesis, while selection is the null hypothesis in methods IX and X. Methods I–VI and X examine the outcome of natural selection, while methods VIII–IX specifically test whether or not its conditions are met. Of the methods that examine the immediate outcome of natural selection, only some (methods IV–VI, IX) can distinguish between selection happening currently in the study area, and selection that occurred there some time in the past. Thus the methods vary considerably in their ability to demonstrate natural selection in the wild, and only methods VI–VIII can provide direct evidence that natural selection is currently happening in a study area. Since the direct methods usually do not address the direct causes of natural selection, it is necessary to utilize several methods to demonstrate natural selection and to show why it occurs. Those who criticize the assumption that natural selection occurs have done so because they have entirely ignored the fact that selection can be detected in ways other than methods IX and X; they have confused the reason for constructing an hypothesis with the method of testing the hypothesis. An investigation of natural selection is not the same as assuming that it exists; the logical structure of hypothesis testing is independent of the source or the nature of the hypothesis.

CHAPTER FOUR

Problems in Detecting Natural Selection

> Nothing is more subject to mistake and disappointment than anticipated judgment concerning the easiness or difficulty of any undertaking, whether we form our opinion from the performances of others, or from abstracted contemplation of the thing to be attempted. . . . In adjusting the probability of success by a previous consideration of the undertaking, we are equally in danger of deceiving ourselves.
>
> Samuel Johnson, *The Rambler*, 18 May 1751

Natural selection is not easy to detect. In addition to the logical and design limitations of the methods discussed in Chapter 3, there are many practical problems. Some of the problems have been known for a long time (Robson and Richards 1936), but others have been appreciated only recently. The problems can be grouped roughly into three kinds, by their effects: (1) lack of detection of natural selection when it is present; (2) apparent detection when it is not present; and (3) misleading demonstrations of natural selection. The first two groups are equivalent to type I and II errors in statistical inference (Sokal and Rohlf 1981): group (1) consists of type II and group (2) consists of type I errors when an absence of selection is the null hypothesis (methods I–VIII); the reverse is true for methods IX and X (see Chapter 3). These three classes grade into one another; for example, any of the problems in (1) and (2) can also yield misleading demonstrations of natural selection. Virtually all of these problems are avoidable with careful and extensive work and with proper sampling design: "forewarned is forearmed." There are no shortcuts in demonstrating natural selection.

This chapter deals only with problems that apply to most or all of the methods. Problems that are specific to one or two methods were discussed in Chapter 3. Readers who are not interested in the practical problems of detecting selection in natural populations should peruse Tables 4.1–4.3, and skip the rest of this chapter, which consists largely of table annotations and examples.

4.1. REASONS FOR LACK OF DETECTION OF NATURAL SELECTION WHEN IT EXISTS

Table 4.1 summarizes problems that obscure the detection of natural selection, and they are elaborated below.

Selection will not be detected if the estimates were made at a place, life-history stage, date, season, or even generation when selection did not occur, or when it occurs only slightly.[1] The ecological conditions may be critical; for example, O'Donald (1971) found variation in fitness only under conditions of intense crowding in *Drosophila*, and Redfield (1973a,b) found heterozygous advantage only in places where *Dendragapus obscurus* grouse were at highest density. In studies involving predation or other mortality factors, selection will not be detected if most individuals have died (Haskins et al. 1961; Krimbas and Tsakas 1971). In addition, if the direction of selection varies at random among generations, the results may be indistinguishable from random genetic drift (Kimura 1954; Wright 1965; Karlin and Lieberman 1974).

Statistical problems may obscure selection. The trait of interest may have a measurement error that is larger than the magnitude of selection (Johnston et al. 1972). Variation in trait value (X) or environment (E) may be too small for selection

[1] Examples are discussed in Berry and Jakobson 1975; Endler 1977; O'Donald 1971, 1973; Redfield 1974; Wicklund 1975.

TABLE 4.1. Some reasons for lack of detection of natural selection, even though it exists

1. Selection did not occur during the attempted estimate.
   a. Intermittent selection.
   b. Seasonal selection.
   c. Inappropriate or unusual conditions.
   d. Direction of selection varies at random.
2. Statistical problems.
   a. Too much measurement error in the trait in question.
   b. Insufficient variation in samples studied to enable the test to work.
   c. Selection too weak to be statistically significant.
   d. Scale of sampling too small relative to neighborhood size.
   e. Statistical tests not very sensitive (method IV).*
3. Insufficient genetic variation in population studied for selection to be effective, although present in other populations.
4. Incorrect assumptions about the mode of selection.
   a. Wrong selection estimator.
   b. Frequency-dependent selection, and system at or near equilibrium.
   c. Presence of equivalent phenotypes; phenotypes that are equivalent with respect to natural selection (Chapter 1, section 1.4.2).
5. Incorrect age class or demographic comparison.
   a. Most selection takes place between age classes that are not measured.
   b. One or more age classes measured are post-reproductive.
   c. Age classes too close together or cohorts not followed long enough.
   d. Age classes too far apart and selection in opposite directions among different age classes.
6. Incomplete or incorrect trait investigated; trait measured is only part of the "unit" actually affected by natural selection.
7. Incomplete identification of genotypes or sexes.
   a. Selection in opposite directions in different sexes or genotypes.
   b. Equivalent phenotype groups present but not distinguished.
8. Overenthusiasm for inductive reasoning.
   a. Aggregation of samples at different times of different years where there is seasonally varying selection.
   b. Aggregation of samples at different places with selection in different directions.
   c. Extrapolation beyond known limits of variation in data.
9. Field behavior minimizes genetic differences found in laboratory.
10. Other, method-specific problems, see especially method IV.

* Problem discussed in Chapter 3 under indicated method.

to be detectable, even with strong selection and large sample sizes. For example, suppose there is a correlation between E and X of $r = 0.6$; then at least nine sample sites would be needed to detect the correlation at the 5% level (or 16 at the 1% level), regardless of their individual sample sizes. Suppose further that E was temperature and the 0.6 correlation was a latitudinal one. If the latitudinal range of sampling were small, this would result in a small observed variation in E, requiring a much larger number of localities to detect the same correlation than if sampling localities were distributed over a larger temperature range; a smaller variance of E, when holding variance of X roughly constant, yields a smaller observed r (Sokal and Rohlf 1981). Similarly, if there were a real but small difference in means of traits between two age classes—if for some reason the total phenotypic variation was small—then it would be very difficult to detect the difference.

In some cases, natural selection may be so weak that it would require a sample size larger than the local population to detect that difference, making detection impossible even with omniscience.[2] In such a case perturbation studies may help because changes in trait distributions are usually greater farther from equilibrium, and also because measurable direct effects of frequency-dependent natural selection are weakest at equilibrium. Even fitness component analysis has weakest statistical power near values predicted by the null hypothesis, and has acceptance regions broad enough to miss some important deviations (Christiansen and Frydenberg 1973).

One must also define the area over which samples are taken; if several samples are taken within a neighborhood, deme, or gene flow unit, then any differences found would be fortuitous, and this random variation may obscure estimates of selection

[2] Details and examples are found in Bantock and Bayley 1973; Christiansen and Frydenberg 1973; Christiansen et al. 1973; Clarke 1975; Hiorns and Harrison 1970; Lewontin 1974; Weir 1979; See also Chapter 7.

as well as trait frequencies (Bantock and Ratsey 1980; Cain 1983; Clarke 1975; Clarke et al. 1978).

Unknown geographic variation in heritability can yield false conclusions. If an attempt is made to detect natural selection in a local population with zero heritability, only phenotypic selection can be observed, and this may wrongly imply that natural selection does not affect that trait (except to those who equate phenotypic selection and natural selection—see Chapter 1). Phenotypic selection could be geographically uniform, with varying heritability; this would yield inexplicable differences in phenotype distributions. There is no reason why nonzero heritabilities should be constant throughout the species range; they apply only to the population for which they wer estimated, for the particular combination of gene frequencies and environmental values at that time (Falconer 1981). This point is often missed in the current rush to measure heritabilities in natural populations. It is invalid and misleading to generalize the outcome and effect of natural selection from a locality with a given heritability to other localities with different heritabilities. This needs further study.

Incorrect assumptions may have been made about the mode of selection. The wrong selection estimator may have been used. For example, a method that estimates the difference in means between selected and unselected individuals will detect directional selection but not other modes, and if other modes contain a strong directional component, the same method will imply that there is only directional selection. Conversely, a method that detects only changes in the variance of the distributions may not detect directional selection. Directional selection for a trait, which is counteracted by opposite directional selection on a phenotypically correlated trait, will result in data that appears to show no evidence of selection unless multivariate methods are used (Lande and Arnold 1983). Composite univariate measures may do no better than to tell us that selection has occurred. Finally, if most selection is sexual, then examining differences in mortality or fecundity may yield insignificant results.

If there is unknown frequency-dependent selection and the system is near equilibrium, then selection should not be detectable; at or near equilibrium the fitnesses of genotypes maintained by frequency-dependent selection are approximately equal (Clarke 1975). Similar kinds of problems may arise if frequency- and density-dependent selection are confounded (Clarke 1975; Wallace 1981), or if fitnesses are assumed to be constant when they are not.

A very serious source of confusion is the unknown presence of equivalent phenotype groups (see section 1.4.2). If comparisons were made within groups, the trait would appear to be selectively neutral; this would not be true of among-group comparisons within habitats, or within-group comparisons among habitats. Unfortunately, the identification of equivalent phenotypes may not be possible without knowing the biological reasons for natural selection. This is a serious problem and warrants further study.

Incorrect or incomplete comparisons may have been made among age classes or cohorts. Selection would not be detected if it took place between age classes that were not included in the samples, or if mostly postreproductive age classes were examined. For example, consider age classes *A*, *B*, *C*, *D*, and *E*, and let selection be most effective between *B* and *C* (Figure 4.1). If only *C*, *D*, and *E* were sampled, then selection may not be detected. If *A*, *B*, *D*, and *E* were sampled, no selection would be detected unless *B* and *D* were compared. If age classes were too close together, then any differences among them may be too small to be detected, even though cumulative selection over more of the lifetime may be highly significant. A similar problem would arise if cohorts were not followed for a long enough time to include a significant proportion of their lifetime. Another potential source of confusion can arise if the

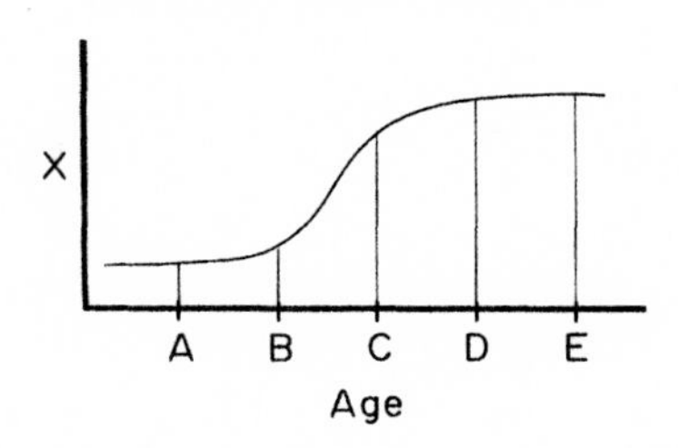

FIGURE 4.1. Average trait of a cohort as a function of age. *X* represents the mean trait frequency (if polymorphic) or the mean or standard deviation of trait value (if a quantitative trait) for the cohort. *A*, *B*, *C*, *D*, and *E* represent age classes or life-history stages through which the cohort passes. They also represent sampling periods for a cohort analysis (method VII). A single sample from a species with overlapping generations would show different *X* for different age classes in an age class comparison (method VIII). For both methods, if only *A* and *B*, or only *C*, *D*, and *E* were sampled, then no evidence for natural selection would be detected, except between *C* and *D* for very large sample sizes; differences in *X* are too small. If all five ages were sampled, or samples happened to include *B* and *C*, then selection would be detected. If a very large number of age classes were sampled between *B* and *C*, then each would yield a very small difference, perhaps too small to be statistically significant.

sampled age classes are too far apart: if selection changes direction between the samples, then the net result may be very small or zero, yielding no apparent selection (Figure 4.2). Possible examples are found in *Mus* (Berry et al. 1978) and *Salmo* (Klar

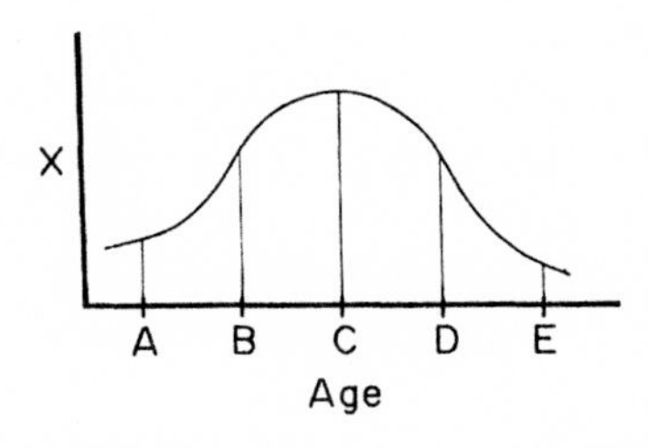

FIGURE 4.2. Average trait of a cohort as a function of age. Axes as in Figure 4.1. A comparison between ages *A* and *E*, or *B* and *D* may not indicate selection. On the other hand, a comparison between *A* and *C* will overestimate the lifetime fitness of large *X* phenotypes, as will a comparison between *C* and *E* for small *X* phenotypes.

et al. 1979a), and a good example of reversing selection direction is found in *Hordeum* (Clegg et al. 1978a).

Incomplete or incorrect traits may have been investigated. The trait measured may be below the level actually affected by natural selection. The problem of what is or is not a trait, and also what level or "unit" (if any) is selected, is a serious problem for philosophical reasons (Ghiselin 1981; Hull 1980; Brandon 1982; Sober 1984) as well as genetic reasons (linkage, epistasis, and pleiotropy; Wright 1978, 1982; Lewontin 1970, 1974). Traits are usually chosen because they are easy to recognize, but this does not mean that they are "recognized" as distinct by natural selection. The adipose fin of certain teleosts is a case in point; it is physically distinct and easy to recognize (second, boneless dorsal fin). Its presence or absence is variable among fish families and genera within families. Hydrodynamic studies (Webb 1975) have shown that it pushes the boundary layer separation zone further toward the tail, improving the efficiency of swimming. However, it only works with certain body and fin proportions at certain velocities. Thus a study of selection on this fin, ignoring effects on the other fins and body shape, may yield false or misleading results. Color patterns provide another example. A study of selection on one color pattern element, ignoring the others, may yield insignificant results since it is the whole pattern rather than the color or size of the single spot that is important in visibility to predators and mates. There may be quite a bit of variation in the details of the spot patterns, but so long as they have a trivial effect on the overall conspicuousness, they make little difference to fitness (Endler 1978, 1980, 1983). Single loci, which are small components of the trait that is influenced by natural selection, may appear to be subject only to genetic drift (Milkman 1982; Kimura and Crow 1978; Lynch 1984). It must be remembered that a single trait may not correctly represent the level or scale of selection.

Incomplete identification may have been made of genotypes or sexes. If genotypes are misidentified, any differences they have may disappear. If selection differs among the sexes, and sexes were not separated, then this may also lead to problems. For example, Berry et al. (1979) found seasonal changes in allele frequencies in opposite directions among male and female mice at three different sites, and opposite selection has also been found between sexes of house sparrows (Johnston and Fleischer 1981; Fleischer and Johnston 1982, 1984). If sexes were combined in these studies there would have been no evidence for selection. As mentioned earlier, unknown equivalent phenotypes are another source of trouble.

Overenthusiasm for inductive reasoning (extrapolation) is a potentially serious problem. If samples are taken at different times in different years, and there is seasonal or other periodic selection, then the net effect of lumping samples may be to cancel any effects of selection; generalization to other seasons is dangerous (Cain 1983; Clarke 1975; Clarke et al. 1978). Similarly, aggregation of samples from different locations in a species with geographically varying selection may eliminate any effects of selection. As shown earlier, an estimate of heritability—and for that matter, any selection differential or coefficient—does not mean that the measured value will remain constant throughout the species range, or in time. Assuming constancy of the conditions for natural selection is one of the last vestiges of typological thinking in biology. Examples are discussed in Cain (1983), Clarke et al. (1978), Endler (1978) and O'Donald (1971).

An excellent example of how an incomplete knowledge of the ecology of a species, together with uncritical generalization, can lead to false conclusions about the prevalence of natural selection, is found in studies of the effects of herbivores on the cyanogenesis polymorphism in both *Lotus corniculatus* and *Trifolium repens*. There was an apparent contradiction in the literature about how well the cyanogenic genotypes of these species protect them against mollusk damage.[3] But different workers were experimenting with different gastropods and plants. *Agriolimax reticulatus* selectively avoids cyanogenic plants of both *Lotus* and *Trifolium*. *Agriolimax caruanae* and *Arion ater* are selective of *Trifolium* and have not been measured on *Lotus*. *Arion hortensis* is selective of *Trifolium* but not of *Lotus*. *Arion subfuscum* is not selective of cyanogenic morphs of either species. *Helix aspersa* is selective of *Trifolium* but not of *Lotus*. Data are heterogeneous among studies for *Agriolimax reticulatus* and *Helix aspersa*, but this may reflect differences in experimental procedure—for example, variations in state of hunger (Whelan 1982) or alternative food items. It should not be a surprise that different species of gastropods will behave differently to different morphs and species of polymorphic cyanogenic plants, and therefore result in different patterns of natural selection, depending upon which herbivores are present (see especially Angseesing 1974). Yet some researchers have assumed that because one species is not selective on cyanogenic morphs of one species of plant, the polymorphism is selectively neutral with respect to the other plant species and to all herbivores. Other researchers have made the reverse assumption, that because some slugs are selective, that all are, and that cyanogenesis is an antiherbivore device in all populations. Both assumptions are good examples of typological thinking; they are clearly wrong because they extrapolate too far.

It is possible that behavior in the field may eliminate any differences that we might expect on the basis of physiology alone. For example, if there is microhabitat or microclimate choice, such as in *Acmaea* limpets (Giesel 1970) or *Cepaea* snails (Clarke et al. 1978; Cain 1983; Jones 1982; Jones and Probert

[3] See Abbott 1977; Angseesing 1974; Angseesing and Angseesing 1973; Barber 1965; Bishop and Korn 1969; Corkill 1952; Crawford-Sidebotham 1972; Daday 1954–1965; Dirzo and Harper 1982a,b; Dritschillo et al. 1979; Ennos 1981; Ellis et al. 1977a–c; Foulds and Grime 1972; Jones 1962–1972; Keymer and Ellis 1978; Whitman 1973.

1980; Tilling 1983), the genotypes may choose the microenvironment in which they are most fit, minimizing or eliminating any measurable effects of natural selection. Habitat choice can have complex and interesting effects on polymorphisms, which are completely different from multiple habitat models with no habitat choice (Rausher 1984; Jones and Probert 1980).

4.2. REASONS FOR APPARENT DETECTION OF SELECTION WHEN IT IS NONEXISTENT

Table 4.2 summarizes problems that lead to false evidence for selection, and they are elaborated below.

Natural selection cannot occur if the trait is not heritable; this is condition *c* for natural selection and was one of the earliest problems to be recognized (Robson and Richards 1936). For example, variation in esterases in terrestrial molluscs may be induced by their diet (Oxford 1977, 1978), and the colors of some grasshoppers such as *Chorthophaga* depend upon the background color at the time of the most recent molt (Otte and Williams 1972). Problems can be quite subtle. For example, variation can actually appear to be genetic for several generations; accidents of feeding success and general health of young red deer of both sexes (*Cervus elaphus*) can affect their survivorship and breeding success, as well as that of their offspring and grandchildren, through maternal care and social effects; yet the variation can have no genetic basis (Clutton-Brock et al. 1982). By the definition given in Chapter 1, this is a borderline case, since there is a significant nongenetic relationship between parents and their offspring for some traits, though these can be regarded as common-environment effects. Another false result can occur when looking for geographic correlations between environmental factors and phenotypes; if there is any phenotypic plasticity, then correlations are expected that have nothing to do with ongoing natural selection. This will also happen for seasonal variation (polyphenism). Such variation can be quite

TABLE 4.2. Some reasons for apparent detection of natural selection, even though it does not exist

1. Trait studied has no heritability.
2. Random effects and statistical problems.
   a. Too many traits studied.
   b. Random changes in traits sampled with single-generation age-class comparisons.
   c. Runs in randomly varying survival.
   d. Chance correlations with environmental gradients (method I).[a]
   e. Unknown trait distribution in absence of selection (method IV).[b]
3. Collection bias.
   a. Genotypic differences in visibility, trapability, etc.
   b. Sampling bias different among different age classes.
   c. Temporal changes in sampling error.
   d. Spatial changes in sampling error.
4. Incomplete knowledge of population structure.
   a. Sampling on a scale larger than random mating unit (neighborhood).
   b. Constant gene and genotype frequencies as a result of large effective population size, large amounts of gene flow, or differential dispersal.
   c. Changes in trait frequencies as a result of immigration in age-class comparisons or perturbation experiments.
   d. Introgression among compared species (method III).[a]
   e. Unknown trait distribution after perturbation or introduction (method VI).[a]
5. Linkage and correlation effects.
   a. Linkage disequilibrium or trait covariation associated with genetic drift and population structure (method IV).[b]
   b. Hitch-hiking and other covariance effects (method IV).[b]
   c. Similar covariance among species due to common genetics (method II).[a]
6. Developmental or Ontogenetic Effects
   a. Growth and age effects.
   b. Developmental canalization affecting trait variance.
7. Other, method-specific problems; see especially methods I, II, IV, VI, IX, X.

[a] Problem discussed in Chapter 3 under indicated method.
[b] Problem discussed in Chapter 3 under indicated method, but the problem can affect almost all methods.

adaptive and predicted by methods IX or X; an example is the seasonal blackening of *Nathalis iole* butterflies, which greatly aids in thermoregulation but is not a genetic polymorphism (Douglas and Grula 1978).

Random effects and statistical problems can arise if we are not careful. If too many traits are studied, and many tests are done, then some will be significant, even though there is no natural selection. A possible example is human blood groups and disease. In a large survey Reed (1968) performed 1,008 significance tests, of which 44 showed significant associations between blood groups and disease at the 5% level. By chance alone we expect about 50 to be significant at the 5% level.

If age-class comparisons are made in too few places or in too few years, and if there is random phenotypic variation, then chance differences among age classes will seem to indicate selection. But such differences will be temporary and have no relevance to natural selection (for example, Berry and Peters 1975). Sampling must be repetitive and intensive to avoid this problem. A good rule of thumb for age-class comparisons is a minimum of six places and/or generations. Consider a comparison between two age classes, yielding a difference (d) between gene frequencies, trait means, or trait variance. If age class variation were random, then we expect an equal number of positive and negative d. A minimum of six comparisons is required to detect a difference from the 1 : 1 expected on the binomial test (Siegel 1956). If more than two age classes are being compared, then the minimum number of samples can be obtained from the multinomial distribution. Prout (1965–1971a,b) suggests eight comparisons for estimates that are derived from between- as well as within-generation data. Random variation is relatively less of a problem in cohort analysis (method VII), but there may still be chance differences in demographic parameters among cohorts. Lininger et al. (1979) and Farewell and Dahlberg (1983) discuss methods for and problems in distinguishing survivorship curves, and more work needs to be done to include other components of selection. Unfortunately, we know almost nothing about the form of the distribution of random variation in demographic parameters in natural populations.

For quantitative traits, it must be certain that there are no environmental changes among generations that affect the expression of traits; if there are, then different age classes sampled at the same time will differ, but the difference will be due to phenotypic plasticity and environmental fluctuations rather than natural selection. This will yield misleading rather than false evidence of selection if heritability is significantly different from zero. This result, as well as problems due to ontogenic changes, can be minimized or avoided by following cohorts through many age classes (Bell 1978; Lande and Arnold 1983) or by long term studies with repeated sampling at the same places.

The problem of random runs in genetic drift, discussed in Chapter 3 (section 3.5.2) may also yield false apparent selection in method VIII; one result would be a consistent difference among age classes. This needs further study.

Biased collection or sampling techniques can yield invalid evidence for selection. For example, certain color morphs will be more difficult to see, and hence be underrepresented in the sample (Wolda 1963; Clarke et al. 1978). Due to differing activities, some genotypes may be more likely to be encountered by a variety of sampling techniques, and certain individuals may actually be attracted to traps (Southwood 1978; Johnson 1981; Tilling 1983). To make matters worse, such differences may change with age (Southwood 1978). If the collecting bias changes between sampling periods, then natural selection will appear to have occurred (Clarke et al. 1978). False apparent selection will also occur if the environment changes while the bias remains constant. For example, suppose green morphs are easier to see on a brown background than brown morphs. Let the background change to greenish between samples, making the browns more conspicuous to the collector, shifting the apparent frequencies. If visual predation by mice is on the basis of tone rather than color, or if there is no natural selection at all, then there will be no shift in the actual morph frequencies, only in apparent (sampled) frequencies, yielding false selection evidence. See Wicklund (1975) for discussion.

Simple changes in sampling error may also result in apparent selection (Clarke et al. 1978). If a particular age class, such as juveniles, is harder to find, it is likely that more effort will be spent looking for them over a wider area than an easier age class, such as adults. Since microgeographic variation is usually present, the sample from the larger area is likely to be more variable than the sample from the smaller area. If the juveniles were taken from the larger area and the adults from the smaller, then this would yield false evidence for stabilizing selection (Clarke 1975; Clarke et al. 1978). This is a significant problem in the early attempts to demonstrate stabilizing selection (Weldon 1901, 1903; DiCesnola 1904; Inger 1942, 1943). Every effort should be made to eliminate, minimize, or correct for sampling bias.

Incomplete knowledge of population structure may cause problems. Sampling of post-reproductive individuals will yield no information about selection (Robson and Richards 1936). Sampling from a scale larger than the neighborhood area can give rise to the Wahlund effect (a deficiency of heterozygotes), which has nothing to do with natural selection (Barndorff-Nielsen 1977; Clarke et al. 1978; Christiansen and Frydenberg 1973; Wallace 1981). If the sampling area is not held constant and there is local variation in trait frequencies on a small scale, then random variation in local frequencies, averaged over different areas at different times, will yield differences in frequencies between subsequent samples that have nothing to do with selection. A similar problem will occur for the variance of trait value.

Gene flow or differential dispersal ability may also yield false conclusions about selection. Constancy of trait values may not be evidence for selection if there is much gene flow from neighboring areas or a large effective population size (minimizing random genetic drift). In addition, changes in trait distributions may come from sudden influx of genes from single areas, or from changes in the pattern of incoming gene flow even though no selection has occurred (Barker and East 1980; Brown 1981). For example, many studies show a positive correlation between human G6PD deficiency and elevation in the Mediterranean region (Tzoneva et al. 1980). This is thought to be caused by the reduction of malaria incidence with elevation in these areas (mosquitos do poorly at higher elevation). However, Brown (1981) found that the correlation in Sardinia was due not to selection by malaria, but to the pattern of historical migrations during the Carthaginian and Roman periods. Another example shows how to prevent these problems. The sudden change in morphology following a drought in Boag and Grant's (1981) study of Galapagos finches could either have resulted from selection for body size by drought, or because the drought resulted in an influx of birds from the very close neighboring island. In this particular case the latter is discounted, since the birds were banded before and after the drought (method VII). A third possibility is differential emigration of smaller birds, perhaps because they would suffer the lack of food earlier than the larger birds and hence would try to find a more salubrious island; but this is also unlikely (Price and Grant, pers. comm. 1983). The possible effects of gene flow and dispersal must be considered in all studies.

Linkage and trait covariance effects may give a false appearance of selection (Chapter 3, Figure 3.2) because they can result from certain forms of population structure. Although discussed for method IV, they can cause apparent selection while using any method. Effects of phenotypic correlations among quantitative traits can be minimized by means of multiple regression (Lande and Arnold 1983), but this will not allow for genetic correlations unless there is also a genetic analysis. The difference between correlations during phenotypic selection and during the genetic response can be made only by looking at both conditions *b* (fitness) and *c* (inheritance) for natural selection (Chapter 1) and provides an additional reason for including condition *c* in the definition of natural selection.

Developmental changes can give misleading or even false evidence for natural selection. If the trait of interest is affected by growth between the sampling periods, then the resulting change in the trait distribution between the samples will affect or even mask any affects of natural selection. In addition, if there is geographic variation in life expectancy (or in age structure), and this is strongly correlated with a geographically varying environmental factor, then the resulting correlation between the environmental factor and the trait distributions will provide false or misleading evidence for selection.

My own work with guppies (*Poecilia reticulata*) illustrates these problems and some solutions (Endler 1978–1983 and unpublished). In guppies both the size and number of colored spots increases (on average) with age in some males. Comparing age classes without accounting for this change would yield false or misleading evidence that males with larger spots are at an advantage. In addition, there is geographic variation in predation intensity and hence life expectancy, so presumably the mean age of fish in high predation areas is smaller than in low predation areas. As a result, the smaller and fewer spots seen in high predation areas could result from low mean age. This applies to the field correlations as well as the field transfer and greenhouse experiments which involve changes in predation intensity followed by changes in color patterns. This problem could also apply to many other morphological traits in many different species. However, and conveniently, the relationship between age and spot patterns in guppies is not linear but decreasing—most of this ontogenetic change occurs within about two weeks or less after sexual maturity. In addition, certain color pattern elements are the last to appear, so mature color patterns are distinguishable from immature patterns. The changes in virtually all mature males are so slight that they can be ignored, except that in about a quarter of the males the color pattern in the caudal (tail) fin continues to change throughout life. Consequently I have restricted my analysis to males with fully mature color patterns, and have either excluded data on the caudal fin patterns, or analyzed them separately. But, even so, ontogenetic effects could account for as much as a quarter of the differences found among populations, and may at least partially account for the extremely rapid response to selection observed in the greenhouse. At the very worst, growth effects cannot explain all of the variation because there is a significant effect of background color pattern (gravel size) within predation treatments, and there are no significant differences in size (and presumably age) within predation treatments. Trait distributions should always be measured at equivalent ontogenetic stages, and if there is any doubt, the causal mechanisms should be tested experimentally.

Note that these problems exist even if the trait has known genetics (as in guppies), so demonstrating mendelian inheritance or significant heritability would be insufficient to avoid these problems. In species where there is a continuous change in the trait with age, all measurements (including genetics) should be made on individuals of identical age and size. Unfortunately, this would require discarding most of the individuals of a field sample. Thus it would be best to do a nonlinear regression of trait value on age and use this to correct for ontogeny in the samples. Unfortunately it is often difficult to age individuals in natural populations, and the relationship itself may change with predation intensity (as in guppies, Reznick and Endler 1982). In such cases it would be especially important to demonstrate the function of the trait and attempt to predict the effect of its variance on fitness (method IX, Chapter 3).

In studies that apparently show stabilizing selection on morphological traits, it is possible that the decrease in variation may be due, at least in part, to developmental canalization that increases with age, reducing the morphological variance with older age classes. An example in which this effect occurs, in addition to stabilizing selection, is Bell's (1974, 1978) work on recently metamorphosed *Triturus* newts. This problem can be minimized by careful developmental studies and the use of cohort analysis (method VII) rather than age class comparisons (method VIII) (examples in Bell 1974, 1978; Arnold and Wade 1984b). Another problem is that directional selection frequently results in a reduction of variance. This reduction can be mitigated by means of a quadratic regression (Lande and Arnold 1983) but this only estimates stabilizing selection components that can appear as quadratic and linear components (see Chapter 6). It is relatively safe to assume that stabilizing selection occurs when there is no evidence of directional selection and phenotypically correlated traits are included in the analysis. It is, of course, possible for both to occur simultaneously.

4.3. REASONS FOR MISLEADING DETECTION OF SELECTION

Many of the factors discussed in Tables 4.1 and 4.2 may also yield misleading estimates. Additional sources of error are summarized in Table 4.3 and discussed below.

An incomplete knowledge of genetics can be quite misleading. The genetic determination of a trait may not be the same in all places. For example, alcohol tolerance in *Drosophila* in vineyards may be due either to the ADH locus or to a quantitative trait where ADH genotype is apparently irrelevant. In places where the quantitative trait is important, there may be no evidence for selection at the ADH locus, though there is clearly selection for ethanol resistance (Briscoe et al. 1975; David

TABLE 4.3. Some reasons that demonstrations of natural selection may be correct but misleading

1. Incomplete knowledge of genetics.
   a. Incomplete identification of genotypes or traits investigated (see Table 4.1).
   b. Generalization of genetic response from one population to another.
   c. Age or condition-dependent heritability.
   d. Hitch-hiking (see Table 4.2).
2. Incomplete knowledge of population structure.
   a. Individuals studied are post-reproductive
   b. Unidentified family units distributed at same scale as sampling unit.
   c. Disassortative mating may yield false heterozygous advantage.
   d. Other sampling and linkage effects (see also Table 4.2).
3. Differences in survival or reproduction in cohort or perturbation studies may be due to general differences among source populations rather than the trait studied. Especially difficult if source populations vary in frequencies of unrecognized sibling species.
4. Geographically or temporally varying selection; estimates may be incomplete or valid for only one time or place (see Tables 4.1 and 4.2).
5. Incomplete knowledge of biology and ecology.
   a. Partial versus lifetime fitness estimates.
   b. Conditions of selection.
   c. Missing important environmental factors.
   d. Countergradient effects.
   e. Injury rate and predation risk.
   f. General ecology.
   g. Extrapolation outside range of measured variables (see Tables 4.1 and 4.2).
6. Incomplete knowledge of dynamics. Equilibrium assumption when not at equilibrium, or vice-versa.
7. Sampling effects may cause apparent frequency-dependent survival and mating.
8. Laboratory experiments with predators.
   a. Predators may be more selective if generally better fed than an average individual in the field.
   b. Apostatic selection and training predator on only one prey item.
9. Many of the problems in Tables 4.1 and 4.2 will also yield misleading estimates of natural selection.
10. Other, method-specific problems; see especially methods IV, VI.

et al. 1979; McKenzie and Parsons 1974; McKenzie and McKechnie 1978). Thus it would be misleading to say that natural selection of alcohol tolerance always effects the ADH locus.

In addition to variation in the mode of genetic determination, there may also be variation in heritability, which may be due to varying age or experimental conditions. Fong (1985) found that heritability varied regularly with age in natural populations of *Gammarus minus* amphipods. This may have resulted from changes in maternal and common environmental components of phenotypic variance, or from genes being turned on and off during ontogeny, which can effect the additive variance components. Measurable heritabilities may change with sampling or ambient conditions because environmentally induced variation may vary with environmental parameters. For example, Tave (1984) found higher heritabilities of dorsal fin ray counts at 25°C than at 19°C in *Poecilia reticulata* guppies and temperature-dependent heritability was also found in *Gasterosteus aculeatus* sticklebacks (Lindsey 1962; Hagen and McPhail 1970). In both of these fish there is a significant relationship between meristic traits and temperature, so the temperature of the experiment affects the outcome. If dependence of phenotypic variation on environmental parameters is ignored, incomplete understanding of selection will result. A similar problem arises from geographically varying heritabilities, as discussed earlier.

If there is a correlation between a trait subject to stabilizing selection and other traits also subject to stabilizing selection, then the degree of stabilizing selection will be overestimated (Lande and Arnold 1983). Simple comparisons of phenotypes, such as between live and dead individuals, can give very misleading impressions about selection, because direct selection of one trait may be opposed by correlated effects of selection of correlated traits, as in the case of *Euschistus variolarius* bugs (Lande and Arnold 1983).

As mentioned earlier, the occurrence of unnoticed equivalent phenotype groups can yield very misleading results. A good example of problems resulting from incomplete recognition of hemoglobin genotypes is found in *Peromyscus maniculatus* mice (Snyder 1983; Nadeau and Baccus 1981, 1983).

Incomplete knowledge of population structure can be misleading. For plants and for animals that live in family groups, it is very important to identify the groups and to sample on a scale larger than a group. If the groups are unknown and of the same spatial scale as the sampling area, then very misleading results will appear (Linhart et al. 1979, 1981). In addition, negative assortative mating (mating of unlike phenotypes) may yield false heterozygote advantage (Clarke 1975), and positive assortative mating may yield false heterozygote disadvantage.

A source of difficulty in field transfers and other forms of perturbation experiments is the confounding of differences between host and source populations with differences between their trait frequency distributions. (The host population is the population into which the introduction or perturbation is made.) Differences in survival or reproduction may be due to general differences between the source and host populations, rather than to the traits of interest. This is especially serious if there is a very large difference between the trait frequencies of source and host. For example, consider a species sampled at two sites, *A* and *B*, differing in mean annual temperature (Figure 4.3). Site *A* has a high frequency of phenotype 1 and site *B* has a low frequency of phenotype 1. The population in site *A* is cold-resistant, while the population in *B* is not. A reciprocal transplant experiment would result in apparent evidence that phenotype 1 is associated with cold resistance, whether or not phenotype 1 had anything to do with it, because the introduced individuals would do more poorly than the natives (Figure 4.3). The example is in terms of mortality, but the effect applies to other components of fitness.

The magnitude of this effect depends upon the differences between the source and host frequency distributions, the percentage of the host population that is replaced by individuals from the source, and selection of various traits. This can be seen by putting the example of Figure 4.3 in a more general form. Let R be the fraction of the host population (B) that is replaced by a random sample of the source population (A). Let p_a and p_b be the trait 1 frequency in A and B, respectively; $q_a = 1 - p_a$, and $q_b = 1 - p_b$. If there is only phenotype-independent mortality (leading to a relative viability of M in A individuals), then the new phenotype frequency is:

$$p' = \frac{MRp_a + (1 - R)p_b}{MR + (1 - R)}. \tag{4.1}$$

If there is only phenotype-dependent mortality (leading to a relative viability of phenotype 1 of W), then the new phenotype

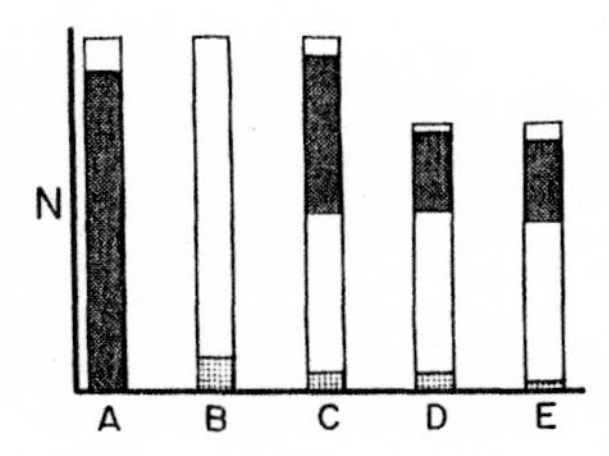

FIGURE 4.3. The effect of trait-specific and trait-nonspecific selection in an introductory experiment in a species with two phenotypes 1 and 2. N is the number of individuals; shaded and hatched portions of the bars represent the numbers of phenotype 1 individuals in the source and host populations, respectively. A, source population, phenotype 1 frequency of $p_a = 0.9$. B, host population before introduction from B, phenotype 1 frequency $p_b = 0.1$. C, population B after 50% ($R = 0.5$) of the population is replaced with a random sample of population A. The new frequency in B is now 0.5. D, population C after 50% mortality ($M = 0.5$), which is independent of phenotype. The new frequency is 0.37, though the frequencies in aliens (a) and natives (n) remain 0.9 and 0.1, respectively. E, population C after 50% mortality ($W = 0.5$) of phenotype 1. The new frequency is now 0.37, but the frequency in aliens is 0.82. A similar result would happen for the reciprocal transplant. Note that the overall frequencies in D and E are similar, but their alien frequencies are different.

frequency is:

$$p' = \frac{WRp_a + (1 - R)p_b}{1 - R(1 - W)p_a}. \tag{4.2}$$

The similarity between the results of equations 4.1 and 4.2 can be quite misleading (Figure 4.3). The effect of both p and R shows how important it is to know exactly how much the population is being perturbed in method VI. The problem can be minimized by examining different genotypes or trait values from both the source and host populations or introducing equal numbers of genotypes in both places (Bantock 1974, 1980; Bantock and Ratsey 1980; Clarke et al. 1978), or by performing backcrosses so that only the traits of interest are introduced (for example, Haskins et al. 1961). It also shows the advantage of cohort analysis (marked individuals) over age-class comparisons (unmarked); in Figure 4.2 the frequency of aliens does not change in D, but it does in E. This problem will be particularly serious if there are differing frequencies of unrecognized sibling species among the experimental populations.

If selection varies temporally or geographically, and only a few estimates are taken, selection estimates will be incomplete and misleading if generalized for the whole species (Figure 4.2). For example, selection on color morphs in *Adalia* beetles may be weak or nonexistent in the winter but strong in the warmer months (Muggleton 1978), and the same may be true for pupal color in *Papilio* butterflies (Wicklund 1975). Selection may vary in direction among different life-history stages; if only one or a few stages were measured, then an estimate of net selection will be too high (examples in Clegg et al. 1978a,b). Stabilizing selection in *Nucella lapillus* dog whelks is strongest in areas of strong wave action, but may be negligible in calmer water (Berry and Crothers 1968, 1970). Geographic variation in selection is also found for color patterns in fish (Endler 1978, 1980, 1982b, 1983) and moths (Kettlewell 1973). Intensive sampling in time and space must be conducted to control for this variation. An estimate of lifetime fitnesses would eliminate some of these problems.

If age-class comparisons are made partially or wholly on post-reproductive individuals, then selection estimates will be quite misleading, unless there is parental care. If sampling occurs after most mortality has taken place, then mortality-related selection will be underestimated. Similarly, if only a few age classes are examined, much of the selection will be missed (Christiansen 1980; Lewontin 1974). A related problem is the appearance of false frequency-dependent effects if samples are taken at the same life stage in different generations, and if selection is incomplete at the time of sampling (Prout 1965–1971a,b).

If only one component of selection is examined, results can be very misleading and poor predictors of future changes in trait frequencies. For example, the color patterns of *Poecilia reticulata* guppies represent a balance between visual predation (favoring crypsis) and sexual selection (favoring conspicuousness). If either one were studied while ignoring the other the geographic variation in color patterns and the results of field manipulations would be inexplicable (Endler 1980). The stage of highest mortality may not necessarily be the stage of strongest selection (Fisher 1930), and this can be especially dangerous if sexual selection occurs and is not measured (see Arnold's 1983a discussion of Howard's 1979 data). A related problem was discussed earlier: if fitness is estimated over only part of the lifetime, then lifetime fitness may be under- or overestimated (Figures 4.1, 4.2).

An incomplete knowledge of the biology and ecology of the species studied can lead to very confusing results, especially when considering traits that affect physiological equilibrium. For example, local selection for maximum growth at particular local mean temperatures, at various elevations and latitudes, can produce apparently counterintuitive results when growth is measured in the laboratory (Berven 1982; Berven et al. 1979): selection may maximize growth rate at a particular local natural temperature, but the difference between the laboratory and natural temperatures, for each elevation, can produce a reverse temperature and growth relationship. Traits that interact with physiology may have complex relationships to environmental factors. For example, Tilling (1983) found that different color morphs of *Cepaea nemoralis* snails heated up differently, and therefore were differentially active at various temperatures. This led to a complex relationship between temperature, insolation, behavior and survivorship, which could not be predicted from any fewer than all these factors. For any study involving several environmental factors and several traits (as in Lande and Arnold 1983), the omission of an important factor can yield very misleading estimates of the importance of the factors that *were* measured; this is a classical problem in the statistical literature for multiple regression (Sokal and Rohlf 1981; Karson 1982). A missing important factor can even reverse the sign of a less important factor.

Correlation studies (method I) or natural perturbations (method VI) may suggest causal relationships, but they may be misleading if the system is not well known. For example the frequency of melanic forms of various British insects increased with the increase of air pollution and decreased with the onset of antipollution laws. This suggests that air pollution affects the frequency of melanics, but how? Two very different cause-effect relationships have been demonstrated: (1) air pollution results in an increase of darkened trees, which, together with visual predation by birds, favors melanics because they are more difficult for birds to see (Kettlewell 1973). (2) Melanics are physiologically more resistant to the direct effects of air pollution (Creed 1971; Kettlewell 1973). In the case of *Adalia* beetles only (2) appears to be true, while both factors affect the frequencies of melanics in moths (Kettlewell 1973). An incomplete knowledge of the biology and ecology, or extrapolation among species, can yield very misleading results.

Some studies have used injury rates in an attempt to estimate risk due to predation. This can be very misleading, because as attack efficiency goes up, relatively fewer individuals will escape a predation event. Minimal injury rates may mean either low predation risk or low escape rate from predators with any degree of risk; a prey may escape undamaged or be eaten (Edmunds 1974; Schoener 1979; Schoener and Schoener 1980; Jaksić and Fuentes 1980; Jaksić and Greene 1984). Predator efficiency as well as attack rate must be known to assess predator risk.

Incomplete knowledge of the ecosystem may also cause problems, as an example from the pesticide literature shows: Central America and India are showing a resurgence of malaria, which is coincident with the great increase in agriculture and insecticide use. The correlation suggests that there is rapid natural selection for insecticide resistance in the mosquitos (Chapin and Wasserstrom 1981), as has actually been demonstrated for other vectors (Brown and Pal 1971). But there are two other explanations. First, the resurgence in India corresponded in time with the conflict in Pakistan and cutoff of foreign aid, suggesting a resurgence due to the reduction of pesticide use. Second, malaria may not not have been extirpated before the increase in insecticide use, but remained in reservoir areas that had not yet been sprayed with insecticides. As expanding agriculture reached these reservoir areas, *Plasmodium* (malaria organism) recolonized the mosquito populations, yielding a resurgence of malaria coincident with agricultural development (Bruce-Chwatt 1981; Curtis 1981). The ambiguity can be eliminated by determining whether or not the mosquitos actually *are* resistant to the pesticides used in the areas of resurgence.

If we assume that a population is at equilibrium when it is not, we can get misleading results. For example, Etges (1984) found that elevational clines for chromosomal inversion types in *Drosophila robusta* changed from positive to negative slope between 1947 and 1981. If selection was estimated using the relationship between a cline's width, gene flow, and the geographic selection gradient (Endler 1977), different and even opposite results would be obtained in different years. If samples were taken from only a few years in sequence, it would be tempting but invalid to extrapolate over the long term, and even the interpretation of mode of selection would be wrong in this case.

Sampling effects, incomplete sampling, incomplete models, and improper indices of selection can result in apparent frequency-dependent survival and mating when selection is not frequency-dependent (Prout 1965–1971a,b; Clarke et al. 1978; Eanes et al. 1977; Goux and Anxolabehere 1980). This can be avoided with proper sampling design and well-thought-out models.

Results of laboratory experiments with predators may not reflect what the same predators do in the field. Predators may be less selective if they are better fed than if they are starved. For example, *Agriolimax caruanae* slugs are less or were not at all food-selective when fed before testing for food preferences (Whelan 1982). Certainly careful experimenters in the laboratory starve the predators for a day or so before the experiment, but the point is they may still be better fed after the experimental "starvation" than they are normally in the field; hence they may be more or less selective in the laboratory than in the field (Muggleton 1978). Very little is known about nutritional levels and requirements of free-ranging predators, and feeding requirements have not been related to food preference experiments. This needs further study.

Another potential problem is found in experiments that examine apostatic (frequency-dependent) predation and train the predator on a particular prey type. First, the wrong or incomplete cues may be used for the training. Song thrushes (*Turdus philomelos*) appear to use color as a more important cue than size (within limits!); if they were trained on a particular size alone, the resulting frequency-dependent effect would be smaller than if trained on both color and size, or one color alone (Harvey et al. 1975). Second, naturally available prey items are very diverse, and it may not be realistic to train predators on only a single prey type; this may artificially increase the measured frequency dependence. In spite of this potential problem, wild bird populations seem to learn the common morphs of a mixed population and will "prefer" these when the frequencies change (Bantock et al. 1975).

4.4. SUMMARY

There are many difficulties in detecting natural selection, and these may be divided into three groups according to their effects (Tables 4.1–4.3): (1) lack of detection of natural selection when it exists; (2) apparent detection when it does not exist; and (3) misleading detection of natural selection. They are not insurmountable with careful and intensive work. The most important problems in detecting natural selection arise out of incomplete knowledge of the ecology and general biology of the species studied, and from incomplete or poorly designed sampling. There is no substitute for careful and intensive field work if one wants to find out what is happening in natural populations.

CHAPTER FIVE

Direct Demonstrations of Natural Selection in the Wild

> Mankind have a great aversion to intellectual labour; but even supposing knowledge to be easily attainable, more people would be content to be ignorant than would take even a little trouble to acquire it.
>
> Samuel Johnson, 1763 (Boswell and Glover 1901, vol. 1, p. 264)

The literature contains a very large number of studies on natural selection. They vary greatly in quality as well as in kinds of species and traits. This chapter summarizes those studies that demonstrate natural selection directly, presents some interesting patterns that arise from the survey, and summarizes the general faults found in most studies.

Of historical interest, the earliest example is an experiment (method VI) that was performed inadvertently on humans, and it actually shows unequivocal evidence for natural selection. In the seventeenth century the Dutch slave traders introduced people from a single area of West Africa (Gold Coast) at random in Surinam and Curaçao. Curaçao has no malaria but Surinam does. The descendants of these people in Surinam today have a high frequency of sickle-cell trait, while those in Curaçao have a very low frequency (Allison 1975).

Another early example is that of Weldon (1899) on crabs. He observed a significant long-term change in frontal breadth of *Carcinus moenas* near an estuary that was becoming progressively more polluted with clay silt and sewage (methods I and

VI). To test the suggested causal relationship (method IX) he kept large numbers of crabs in running clear sea water, sea water with clay silt, and sea water with silt and sewage. He showed, by following cohorts (method VII), that crabs with smaller frontal breadth survived better than the others, and the major source of mortality was silt. The reason is problematic but might be a function of the relationship between frontal breadth and efficiency of water circulation past the gills; but this was never tested. Although there are major problems with the design, including inadequate controls (Robson and Richards 1936), this is a good early example of an attempt to use a laboratory study to test a hypothesis suggested by field correlations. In their book Robson and Richards summarize and critique many of the other early studies of natural selection.

5.1. CHARACTERISTICS OF DEMONSTRATIVE STUDIES

The number of studies actually demonstrating natural selection is a small subset of the literature on natural populations, as might be expected from the problems discussed in Chapters 3 and 4. Yet in spite of the problems, there is a remarkable diversity of direct demonstrations of natural selection in animals and plants, for many different kinds of traits. These studies are summarized in Table 5.1.

The criteria for inclusion in Table 5.1 are as follows: (1) The studies should involve natural populations, or in some cases (such as *Hordeum*), experimental field populations that have been exposed to uncontrolled natural environmental conditions for many generations. (2) Studies of natural selection in the laboratory or greenhouse have been ignored because of their unknown relevance to the evolution of the organisms concerned. (3) Studies were included if they utilized the direct methods (VI–VIII) and they minimized or eliminated the problems discussed in Chapters 3 and 4. This resulted in the elimination of many classical papers on selection. (4) The studies provide evidence for conditions *a*, *b*, and *c* for natural selection (Chapter 1), or were good perturbation studies (method VI). In many cases a study was included if condition *c* (inheritance) was strongly inferred from other species. No special attempt was made to track down papers exclusively dealing with the genetics of the species in the table; hence some species traits that are not marked as having known genetics may in fact be known. Thus all studies in Table 5.1 demonstrate phenotypic selection, and most demonstrate all components of natural selection. Some superb field experiments on color patterns, such as those by Blest (1963), Benson (1972), and others summarized in Ford (1975) and Endler (1978), were not included because they involved artificial changes in the phenotype or were cases in which condition *c* was clearly not true. These are clear cases of phenotypic selection and suggest how natural selection could work if the variation were inherited. (5) Some borderline cases using age-class comparisons (method VIII) were included—for example, the much-analyzed Bumpus (1899) data on *Passer*. These studies are very weak because they include only one or very few samples. As mentioned in Chapter 4, the comparison of age classes in a few samples does not eliminate the serious problem of the chance that random differences between age classes may appear as evidence for natural selection. However, since some of the marginal cases resulted from rare and catastrophic events (e.g., storms), they may indicate something about properly verified natural selection and should encourage more extensive work on rare events. The more critical reader can recognize and exclude these cases by reference to the number of localities and generations over which age-class comparisons were made, which is given for all studies using only method VIII. A good rule is a minimum of six places and/or generations (see Chapter 4). Elimination of these studies has no effect on the patterns discussed here or in Chapter 7.

TABLE 5.1. Direct demonstrations of natural selection

Species	Traits Examined	Method of Demonstration	Selective Agent	References
ANIMALS:				
COLENTERATA				
Acropora sp.	morphology[q]: m,v	VII, VI (introduction)	unknown, physical factors	Potts 1978, 1984
MOLLUSKA				
Bivalvia:				
*Agerostrea mesenterica**	3 morphological traits: m,v	VIII (fossil)	unknown	Sambol & Finks 1977
*Cardium edulae**	ribs per shell: m,v	VIII (live/dead): single samples	wave action?	Palenzona et al. 1971
*Modiolus demissus**	allozymes: *lap*, *to*	VIII: 2 loc., 1 gen.	unknown	Schopf et al. 1975, Koehn et al. 1973
Mytilus californicus	*lap* allozyme	VIII: many loc., 4 gen.	unknown	Levinton & Koehn 1976
*M. edulis**	*lap* allozyme	VIII: cline, 3 gen.	salinity, temperature, and physiology	Koehn et al. 1980, Koehn & Immerman 1981, Koehn & Siebenaller 1981, Hilbish & Koehn 1985

(*continued*)

TABLE 5.1. (*continued*)

Species	Traits Examined	Method of Demonstration	Selective Agent	References
	shell color pattern	VIII: 3 loc., 1 gen.	heat load?	Mitton 1977
Gastropoda:				
*Acmaea digitalis**	shell color pattern	VII, VIII	bird predation	Giesel 1970
*Cepaea nemoralis** *and/or C. hortensis**	shell color[p] & banding (2)[p]	VII, VIII (live/dead)	microclimate, heat balance; overcrowding?	Bantock & Ratsey 1980, Bantock 1974, 1980, Richardson 1974, Tilling 1983, Heath 1975, Emberton & Bradbury 1963
		VII, VIII (live/dead), VI (following background changes)	bird predation	Harvey et al. 1975, Sheppard 1951a, Cain & Sheppard 1954, Wolda 1963, Carter 1968, Cain & Curry 1968, Hakkinen & Kopenen 1982, Clarke & Murray 1962
	Shell size[q]: m	VII	fecundity & unknown	Wolda 1963, 1967
		VII, VI (field cages)	microclimate,	Knights 1979
		VII, VIII (live/dead)	bird predation	Bantock & Bayley 1973, Bantock et al. 1975

		VII	fecundity & unknown	Wolda 1963, 1967
*Littorina picta**	shell surface sculpturing[p,q]: m	VII, VI (introduction)	wave action	Struhsaker 1968
*Nucella (Thais) lapillus**	shell size & shape[(q)]: m,v	VIII: 5 loc., 1 gen.	wave action; predation?	Berry & Crothers 1968, 1970
Tegula funebralis	choice of resting site, escape behavior[(q)]: m	VI, VII	*Octopus* & *Pisaster* predation	Fawcett 1984
CRUSTACEA				
Diaptomus spp. & Calanoids	Carotenoid pigments	VII (direct comparison)	predation, heat balance, & metabolic rate	Byron 1982
Balanus amphitrite	7 allozymes[p]	VIII 6 loc., 1 gen.	temperature?	Nevo et al. 1977
*Sphaeroma rugicauda**	color pattern[p]	VII (lab), correlated with field changes in frequency; VIII: 2 loc., 3 gen.	temperature, growth rate interaction	Bishop 1969, Heath 1974, Khazaeli & Heath 1979
ARACHNIDA				
Tetranychus urticae	host plant resistance: m	VI (introduction on new host, lab)	host plant toxicity	Gould 1979

(continued)

TABLE 5.1. *(continued)*

Species	Traits Examined	Method of Demonstration	Selective Agent	References
INSECTA				
Orthoptera: *Mantis religiosa*	color form	VII	bird predation	DiCesnola 1904
Psocoptera: *Mesopsocus unipunctatus**	abdomen[p] & head/thorax[p] color	VII (field & aviary; significant only in aviary)	predation by *Parus major*	Popescu et al. 1979
Homoptera: *Nuculaspis californicus*	host plant resistance[q]: m	VI, VII (including direct comparison)	host plant toxicity	Edmunds & Alstad 1978, 1981, Alstad & Edmunds 1983a,b
Philaenus spumarius	color pattern[p]	VI (introduction, also environmental change)	unknown; background? vole predation?	Halkka & Raatikainen 1975, Halkka et al. 1975
Hemiptera: *Euschistus variolarius*	2 morphological traits: m,v	VIII (live/dead), 1 loc., 1 gen.	storm?	Lande & Arnold 1983
Uroleucon caligatum	host plant resistance[(q)]: m	VII	host plant (*Solidago* spp.) genotype	Moran 1981

Mecoptera:				
Harpobittacus nigriceps	forewing length: m,v	VII	prey size and mate choice (both sexes), fecundity; viability?	Thornhill 1983
Hylobittacus apicalis	prey size: m,v	VII	mating success (female choice)	Thornhill 1984
Coleoptera				
*Adalia bipunctata**	melanism[p]	VI (+, − changes in environment)	heat balance, physiology, predation, heat balance?	Creed 1971, Muggleton et al. 1975, Majerus et al. 1982a,b
	mating preference[q]	VIII (mated & unmated pairs)	mating ability? activity?	Muggleton 1979, Majerus et al. 1982a,b, O'Donald et al. 1984
*Chauliognathus pennsylvanicus**	5 morphological traits: m, (v)	VIII (mated & unmated pairs)	mating ability?	McCauley & Wade 1978, McLain 1982, Mason 1980
*Tetraopes tetraophthalmus**	*pgm* allozyme	VIII (mated & unmated pairs)	unknown	Eanes et al. 1977
	wing length, body size: v	VIII (mated & random pairs)	mating ability?	Mason 1964, Scheiring 1977, Eanes et al. 1977, McCauley 1979

(*continued*)

Table 5.1. *(continued)*

Species	Traits Examined	Method of Demonstration	Selective Agent	References
Lepidoptera:				
Miscellaneous species	melanism[p]	V, VI (+, − changes in visual background)	bird predation	Kettlewell 1973
Moths:				
Allophyes oxyacanthae, *Diurnea fagella*	melanism[p]	VII	bird predation	Steward 1977
*Amathes glareosa**, *Lasiocampa quercus**	melanism[p]	VII, VIII	bird predation, temperature? physiology?	Kettlewell 1973, Kettlewell et al. 1969, Kettlewell & Berry 1969, Endler 1977, C. Johnson 1976
*Biston betularia**	melanism[p]	VII, VIII	bird predation, physiology	Kettlewell 1958, 1973, Steward 1977, Bishop et al. 1978, Murray et al. 1980, Cook & Mani 1980, Creed et al. 1980, Endler 1977, C. Johnson 1976, Clarke & Sheppard 1966

*Gonodontis bidentata**	melanism^{p}	VII	bird predation, temperature?	Bishop & Harper 1970, Kettlewell 1973
Lycia hirtaria	melanism^{p}	VII	bird predation	Kettlewell 1973
*Panaxia dominula**	wing color pattern^{p}	VI (introduction), VII (including direct observation)	differences in fertility, mate preference, bird predation	Sheppard 1951b, Sheppard & Cook 1962, Ford & Sheppard 1969, Ford 1975
Phigalia pilosaria (=*pedaria*)	melanism^{p}	VII	bird predation, physiology	Lees 1971, Murray et al. 1980, Conroy & Bishop 1980
*Philosamia cynthia**	pupa size & shape: m,v	VII, VIII (live/dead)	physiology, morphology	Crampton 1904
*Phryganidia californica**	wing size: m,v	VIII (mated & random pairs), 2 days	female choice & male mating ability?	Mason 1969
Butterflies:				
Colias, 8 species	melanism^{p}	VII (physiology, biophysics)	heat balance	Watt 1968
Colias philodice	larva color^{p}	VII	bird predation	Gerould 1921, 1926, Hoffman & Watt 1974

(*continued*)

TABLE 5.1. *(continued)*

Species	Traits Examined	Method of Demonstration	Selective Agent	References
Colias philodice & *C. eurythene*	*pgi*^{p}, *α-gpdH*^{p}	VII, IX	temperature & flying period due to enzymatic efficiency, survivorship, & fecundity	Watt 1977, 1983, Watt et al. 1983, Sactor 1975, McDonald 1983, G. Johnson 1976
Hypolimnas misippus	hind-wing color	VII	mimicry of *Danaus chrysippus*	Edmunds 1966
Euphydryas editha	color pattern elements^{q}: m	VIII: 2 loc., 6 gen.	unknown	Ehrlich & Mason 1966
*Maniola jurtina**	spot number^{p,q}: m	VIII: 1 loc., frequency constant, 4 gen.	predation by *Apanteles tetricus* (?), microclimate?	Dowdswell 1961, 1962, O'Donald 1971, 1973, Ford 1975
Diptera:				
Drosophila buzzatii	allozymes: *Est*-2, *Pyr*, *Adh*-1	VI	unknown	Barker & East 1980
D. melanogaster	ebony body color^{p}	VI (introduction)	unknown	Gordon 1935
	adp^{p}, alcohol tolerance^{q}: m	VI (known changes in alcohol concentration in the environment)	alcohol	McKenzie & McKechnie 1978, McKenzie & Parsons 1974

*Lucilia cuprina**	resistance to dieldrin[p]	VII	toxicity	Whitton et al. 1980
Hymenoptera:				
*Centris pallida**	male size: m	VII	mating success (male-male competition)	Alcock 1979, 1984
*Pogonomyrmex desertorum**	male size: m, (v)	VIII (mated & random pairs): 2 gen.	female choice? male mating ability?	Davidson 1982
FISH				
Cottus bairdi	male size: m	VIII (mated & random pairs)	defense of nest sites, abandonment & hatching success	Downhower & Brown 1980
Fundulus heteroclitus	*ldh* allozymes[p]	VII (direct observation)	biochemical properties affecting O_2 balance, swimming speed	DiMichelle & Powers 1982a,b, Place & Powers 1979, Powers et al. 1979
	ldh-b[p]	VII (direct observation)	time to hatching, hatching temperature, O_2 concentration	DiMichelle & Powers 1984
*Gasterosteus aculeatus**	spines,[q] gill raker number,[q] lateral plates[p,q] m, v	VIII (including live/dead), VII	predation: fish, birds, snakes	Gross 1978, Bell & Haglund 1978, Hagen & Gilbertson 1972, 1973a,b, Hagen 1973, Moodie 1972, MacLean 1980

(*continued*)

TABLE 5.1. (*continued*)

Species	Traits Examined	Method of Demonstration	Selective Agent	References
	male color[p]	VII (direct observation)	nesting success, competition for nest sites with *Novumbra hubbsi*	Hagen et al. 1980, McPhail 1969
*Katsuwonus pelamis**	transferrin[p]	VIII: 4 loc., 1 gen.	unknown	Fujino & Kang 1968
Oncorhynchus nerka	allozymes: *ldh-b pgm-1*	VII, VIII	unknown	Kirpichnikov & Ivanova 1977
	body size[q]: m	VIII, VI (overfishing)	predation by bears, man	Konovalov & Shevlyakov 1978
Phalloceros caudimaculatus	color pattern elements: 1,[(p)] 2[(q)]: m,v	VIII: 29 loc., 1 gen.	fish predation	Endler 1982b
Poecilia reticulata	color pattern elements >15,[p] >3,[q] body shape[q]: m,v	VII, VIII, VI (introduction), IX	predation & sexual selection	Endler 1978, 1980, 1983 & unpublished
	5 life history parameters[q]: m	VI (introduction)	predation	Reznick & Endler 1982, Reznick 1982

*Zoarces viviparus**	allozyme *Est*-3	VIII (fitness component analysis), many generations	unknown	Christiansen 1977, 1980, Christiansen et al. 1973a,b, 1974a,b, 1977a,b, Christiansen & Frydenberg 1973, 1976, Ostergaard & Christiansen 1981
AMPHIBIANS				
*Triturus vulgaris**	9 morphological traits[q]: m,v	VII, VIII	predation by shrews (?)	G. Bell 1974, 1978
Bufo boreas	9 allozymes[p]	VIII: 1 loc., 3 gen.	unknown	Samollow 1980
*Hyla regilla**	color variants	VIII: 1 loc., 3 gen.	predation? temperature?	Jameson & Pequegnat 1971
*Rana catesbeiana**	male size: m,v	VII (including direct observation)	number of mates, hatching success	Howard 1979, Arnold 1983a, Arnold & Wade 1984b
R. pipiens complex*	percent abnormal	VIII: 1 loc., 2 gen.	predation?	Merrell 1969
	Burnsii color pattern[p]	VIII: 1 loc., 6 gen.	predation? winter cold?	Merrell & Rodell 1968
*R. sylvatica**	male size: m,v	VIII (mated & random pairs)	mate choice; male competition?	Howard 1980, Howard & Kluge 1985

(*continued*)

TABLE 5.1. (*continued*)

Species	Traits Examined	Method of Demonstration	Selective Agent	References
REPTILES				
*Nerodia (Natrix) sipedon**	color pattern[q]: m,v	VIII: 8 loc., 1 gen.	predation?	Beatson 1976, Camin & Ehrlich 1958, Ehrlich & Camin 1960
*Uta stansburiana**	4 morphological traits[q]: m,v	VII, VIII	unknown	Fox 1975
	hatching size: m	VII	feeding ability? territoriality? predation	Ferguson & Fox 1984
BIRDS				
Anser caerulescens	hatching date[q]: m,v	VII (including direct observation)	timing of fledging	Cooke & Findlay 1982
Domestic chickens	time to achieve homeothermy after hatching[q]	VII (direct observation)	infection by *Salmonella pullorum*	Hutt & Crawford 1960
	resistance to *Salmonella*[q]: m	VI (+, − change in environment)	infection by *Salmonella gallinarum*	Hutt 1963
Domestic ducks	egg size[q] & weight[q]: m, v	VIII (live/dead)	physiology?	Rendel 1943, Haldane 1954

*Dendragapus obscurus**	*Ng* gene[p]	VIII: 8 loc., 5 gen.	density? winter cold?	Redfield 1973a,b, 1974
Geospiza conirostris	3 morphological traits[q]: m,v	VIII: 1 loc., 4 gen.	foraging ability	Grant 1985
*G. fortis**	3 morphological traits[q]: m,v	VII, VIII	foraging ability and food supply	Grant et al. 1976, Boag & Grant 1981, Grant & Price 1981, Schluter et al. 1985
	body weight[q], beak depth[q]: m (juv. & adult)	VII	food supply	Price & Grant 1984, Price et al. 1984
	body size[q], extent of adult plumage, size of territory: m		mate choice	Price 1984
G. scandens	3 morphological traits[q]: m,v	VIII: 2 loc., 1 gen.	foraging ability	Grant et al. 1976, Grant & Price 1981
Molothrus ater	2 morphological traits[(q)]: m	VIII (live/dead), 1 loc., 1 gen.	thermal balance	Johnson et al. 1980
*Parus major** & *P. caeruleus*	fledgling weight, brood size: m,v	VII	unknown, physiology?	Perrins 1965

(*continued*)

TABLE 5.1. *(continued)*

Species	Traits Examined	Method of Demonstration	Selective Agent	References
*Passer domesticus**	3 morphological traits[q]: m,v	VIII (live/dead): 1 loc., 1 gen.	cold shock	Bumpus 1899, Johnston et al. 1972, O'Donald 1973, Grant 1972b, Manley 1976, Lande and Arnold 1983
		VIII: 19 loc., 1 gen.	winter cold?	Lowther 1977, Johnston & Fleischer 1981, Fleischer & Johnston 1982, 1984
		VI (introduction with known source)	thermal balance?	Johnston & Selander 1971, 1973, Baker 1980, Calhoun 1947
*Stercorarius parasiticus**	body color[p]	VII (including direct observation of mating and nesting behavior)	differences in behavior, time to hatching, age at breeding	O'Donald & Davis 1959, 1975, Berry & Davis 1970, O'Donald et al. 1974
	clutch size, breeding date	VII (direct observation)	nest site & food availability?	O'Donald 1972
Sturnus vulgaris	clutch size: v	VII (success rate, weight of young)	parents' ability to feed young	Crossner 1977

MAMMALS				
Domestic cattle	coat color[p,q]	VII (direct comparison)	heat load; fecundity? mortality?	Finch & Western 1977
	transferrins[p]	VII (direct comparison)	fertility, fetal mortality, maternal-fetal incompatibility	Ashton 1959, 1965
*Geomys bursarius major**	chromosome number	VIII: 1 loc., 2 gen.	unknown	Patton et al. 1980, Baker et al. 1983
*Homo sapiens**	tooth size[q]: m,v	VIII: 1 loc., subfossil	unknown	Perzigian 1975
	birth weight,[q] gestation time: m,v	VII, VIII (live/dead)	physiology, unknown	Karn & Penrose 1951, Van Valen & Mellin 1967, Blurton-Jones 1978, Ulizzi et al. 1981
	height[q]: v	VII (direct comparison)	fertility, mating success	Shami & Tahir 1979, Cavalli-Sforza & Bodmer 1971
	body shape: v	VII (direct comparison)	fertility	Clark & Spuhler 1959
	haemoglobin S[p]	VII, VIII (along a cline)	intrinsic, malaria	Allison 1954, 1955, 1975, Ingram 1957, Bodmer & Cavalli-Sforza 1970, Templeton 1982
	g6pd[p]	VII VIII (along a cline)	intrinsic, malaria	Allison 1975, Allison & Clyde 1969, Luzatto et al. 1969, Tzoneva et al. 1980, Brown 1981

(*continued*)

TABLE 5.1. (*continued*)

Species	Traits Examined	Method of Demonstration	Selective Agent	References
*Mus musculus**	coat color[p]	VI (changes in cat density)	domestic cat predation	Brown 1965
	allozymes: *mod*, *got*, *est*	VIII: 3 islands: 1 gen.; 1 island: 2 loc., 2 gen.	unknown	Berry et al. 1978, 1979, Berry & Murphy 1970, Bellamy et al. 1973
	haemoglobin	VII, VIII: 2 loc., 2 gen.; 2 loc., 1 gen.	unknown	Berry & Peters 1975, Myers 1974, Berry & Murphy 1970, Bellamy et al. 1973
	resistance to *Salmonella*: m	VI (+ and − change in environment)	*Salmonella typhimurium*	Gowen 1963
Neotoma lepida	abdominal color[(q)]: m	VIII: 5 loc., 1 gen., along a cline	unknown	Lieberman & Lieberman 1970
Oryctolagus cuniculus	resistance to Myxomatosis[q]: m	VI (introduction)	Myxoma virus	Fenner & Ratcliffe 1965
*Peromyscus leucopus**	tooth size: m,v (body size)	VII (direct comparison of fertilities)	fertility differences	Leamy 1978
P. maniculatus	3 allozymes,[p] albumin[p] & haemoglobin[p]	VIII (fitness component analysis): 1 gen.	unknown	Nadeau & Baccus 1981, 1983, Snyder 1983

*Rattus rattus**	warfarin resistance[p]	VI (use of warfarin)	warfarin	Bishop et al. 1977, Partridge 1979, Greaves & Ayers 1967, 1969
Various fossil mammals, including:	tooth size[(q)]:	VIII (fossil): 1–3 loc.	unknown	Kurten 1953–58, Van Valen 1963a, 1965a, Marcus 1969
*Merychippus primus**	m,v			
*Ursus spelaeus**	m			
*U. arctos**	v			
PLANTS				
FILICINAE:				
Pteridium aquilinum	cyanogenesis: m	VII	sheep & deer grazing	Cooper-Driver & Swain 1976
GYMNOSPERMAE:				
Pinus ponderosa	monoterpenes[p,q]: m	VII (direct comparison)	infestation by *Dendroctonus* spp., cone predation by squirrels	Sturgeon 1979, Farrentinos et al. 1981
	toxicity[q] to herbivores: m	VII (including direct comparison)	infestation by *Nuculaspis californicus*	Edmunds & Alstad 1978, 1981, Alstad & Edmunds 1983a,b
	2 allozymes: *per, est*	VIII, VII (direct comparison)	fertility, growth rate, survival	Linhart et al. 1979, 1981

(*continued*)

TABLE 5.1. (*continued*)

Species	Traits Examined	Method of Demonstration	Selective Agent	References
ANGIOSPERMAE: Monocotyledons: Miscellaneous cereals	resistance to insects[p,q]: m	VII (including direct comparison)	insect damage	Gallun et al. 1975
Agrostis canina	heavy metal tolerance[q]: m	VI (addition of metal fence)	metal ions	Bradshaw et al. 1965
A. stolonifera	stolon length,[q] leaf length[q]: m	VIII: 2 loc., 1 gen.	soil moisture? exposure?	Aston & Bradshaw 1966
*A. tenuis**	heavy metal tolerance[q]: m,(v)	VII, VIII (along a cline), VI (introduction)	metal ions, low soil nutrients, competition	McNeilly & Bradshaw 1968, Hickey & McNeilly 1975, Gartside & McNeilly 1974, McNeilly 1968, 1979, Karataglis 1980a,b
*Anthoxanthum odoratum**	5 morphological[q] & 2 physiological[q] traits: m	VI (introduction, known values)	soil conditions	Snaydon 1970, Snaydon & Davies 1971

	heavy metal tolerance[q]: m,v	VI (introduction, known values)	heavy metal ions, low soil nutrients, competition	Hickey & McNeilly 1975, McNeilly 1979
*Avena barbata**	color *BB*,[p] hairiness *Ls*[p]	VI (introduction)	unknown	Jain & Rai 1980
	3 allozymes,[p] *Ls*,[p] *BB*[p]	VII, VI (introduction & comparison with original population)	soil moisture?	Clegg & Allard 1972, 1973, Hamrick & Allard 1975, Allard, Babbel et al. 1972
	allozymes: 3 *est* loci	VIII: 1 loc., 2 gen.	unknown, some fertility effects	Clegg & Allard 1973
	outcrossing: m,v	VIII (fitness component analysis)	unknown	Allard et al. 1977, Clegg et al. 1978b
A. byzantium & *A. sativa*	3 morphological traits[q]: m	VI (introduction)	microclimate	Marshall 1976
*A. fatua**	lemma color (*B*/*b*)[p]	VII (germinated & dormant seeds, fecundity tests)	unknown, related to density	Jain & Rai 1977
Bromus mollis	*adh* allozyme[p]	VII (germination tests)	temperature	Brown 1979, Brown et al. 1976
Festuca ovina	heavy metal tolerance[q]: m	VI (addition of metal fence)	metal ions	Bradshaw et al. 1965

(continued)

TABLE 5.1. *(continued)*

Species	Traits Examined	Method of Demonstration	Selective Agent	References
*Hordeum vulgare**	5 polymorphic,[p] 3 morphological traits[q]: m	VI (introduction)	climate & soil?	Choo et al. 1980a,b, Jain & Allard 1960, Allard & Jain 1962
	3 polymorphic,[p] 2 morphological traits[q]: m	VI (introduction)	climate & harvest date	Sing & Johnson 1969
	3 allozymes[p]	VI (introduction)	unknown	Allard, Kahler et al. 1972, Weir, Allard et al. 1972, 1974
	esterases[p]	VIII: 1 loc., 3 widely separated gen.	unknown	Kahler, Clegg et al. 1975
	outcrossing[p]	VI (introduction)	unknown	Kahler, Clegg et al. 1975
	3 allozymes & morphological traits[(p,q)]: m	VIII (fitness component analysis)	soil moisture? climate? fecundity?	Clegg et al. 1978a,b, Kahler et al. 1975
Phalaris arundinacea	alkaloid concentration[q]: m	VII (grazed & ungrazed plants, palatability)	grazing by voles, rabbits, sheep, cattle	Williams et al. 1971, Simonds & Marten 1971, Asay et al. 1968, Marten et al. 1973, Kendall & Sherwood 1975

*Plantago lanceolata**	heavy metal tolerance[q]: m	VII, VI (introduction)	metal ions, low soil nutrients, competition	Hickey & McNeilly 1975, McNeilly 1979
Sorghum vulgare	heading time,[q] culm length[q]: m	VI (introduction)	geographical variation in climate & soils	Imai & Gomez 1979
Dicotyledons:				
Delphinium nelsonii	flower color	VII (direct observation)	pollinator efficiency, seed set	Waser & Price 1981, 1983
Eremocarpus setigerus	seed color pattern,[q] palatability: m	VII	predation by mourning doves, germination requirements	Cook et al. 1971
*Eucalyptus urnigera**	glaucousness[p]	VIII (along a cline)	water balance?	Barber & Jackson 1957, Barber 1955, Endler 1977
Glycine max	nematode resistance[p]	VI (introduction)	fecundity differences	Luedders & Duclos 1978
	maturity date[q]: m	VI (introduction)	harvesting time	Luedders 1978
Gossypium barbadense	pure strains[q]: m	VI (introduction)	climate & soil?	Feaster et al. 1980
G. hirsutum	8 morphological traits[(p,q)]: m	VI (introduction)	climate & soil?	Quisenberry et al. 1978

(*continued*)

Table 5.1. (*continued*)

Species	Traits Examined	Method of Demonstration	Selective Agent	References
	gossypiol concentration[q]: m	VII (direct comparison)	infestation by *Heliothis zea*	Lukefahr & Houghtaling 1969
Liatris cylindracea	allozyme *est*-3	VIII: one sample, but ages 2–44 years	unknown; some fertility effects	Schall & Levin 1976
*Lotus corniculatus**	cyanogenesis[p] (two loci)	VII, VI (introduction)	predation by *Microtus*, many gastropod spp., rabbits?	D. Jones 1962, 1966, 1972, Ellis et al. 1977a,b,c, Keymer & Ellis 1978, Crawford-Sidebotham 1972, Compton et al. 1983
		VII, VI (introduction)	drought, salt?, soil moisture	D. Jones 1970, 1972, Keymer & Ellis 1978, Ellis et al. 1977a,b, Foulds & Young 1977, Foulds 1977, Abbott 1977
Lupinus bakeri, *L. caudatus*, *L. floribundus*	alkaloids[(p,q)]: m	VII (direct comparison)	infestation of inflorescences by *Glaucopsyche lygdamus*	Dolinger et al. 1973

*L. nanus**	flower color^p	VII (direct comparison)	fertility, seed set, & viability effects	Harding 1970
*Medicago polymorpha**	spine length D/d^p	VII (germinated & dormant seed, fecundity tests)	unknown; related to moisture?	Jain & Rai 1977
*Mimulus guttatus**	copper tolerance^q: m	VI (habitat change: mining)	copper poisoning	Allen & Sheppard 1971
	chlorophyll deficiency W/w^p	VII (direct observation), VIII	differences in fertility & rhizome growth	Kiang & Libby 1972
*Phlox pilosa**	corolla color^q: m	VII (direct observation)	affects pollen contamination rate by *Phlox glaberrina*, hence seed set	Levin 1968, Levin & Kerster 1967, 1970
P. drummondi	corolla color^q & shape^q: m	VII (direct observation)	affects pollen contamination rate by *Phlox glaberrina*, hence seed set	Levin 1969
Ricinus communis	Bloom on leaves,^{p} stems,^{p} petioles^p	VII, VI (introduction)	sunlight, fecundity	Harland 1947

(*continued*)

TABLE 5.1. (*continued*)

Species	Traits Examined	Method of Demonstration	Selective Agent	References
*Rumex acetosa**	heavy metal tolerance^{q}: m	VII, VI (introduction)	metal ions, low soil nutrients	Hickey & McNeilly 1975, McNeilly 1979
*Salvia columbariae**	seed color	VII	predation by mammals? birds	Brayton & Capon 1980
Senecio vulgaris	ray floret morphology^{p}	VII (direct observation)	fecundity effects; due to pollinator efficiency?	Oxford & Andrews 1977, Stace 1977
Solidago altissima	resistance to aphids^{q}: m	VI (direct observation)	*Uroleucon caligatum* aphids	Moran 1981
Spergula arvensis	seed coat papillae^{p}	VII	germination time, & temperature, fertility	New 1958, 1959, 1978
Trifolium hirtum	3 allozymes,$^{(p)}$ 3 polymorphic,^{p} 6 morphological traits: m	VI (introduction & comparison with original pop.)	unknown	Martins & Jain 1980, Jain & Martins 1979
*T. repens**	cyanogenesis^{p}: (two loci)	VII, VIII (along a cline), VI (introduction)	temperature, fertility, viability effects	Daday 1954a,b, 1958, 1965

		VII, VI (introduction)	soil moisture, salt?, drought	Foulds & Grime 1972, Foulds & Young 1977, Foulds 1977
		VII	predation by cattle, deer, rabbits (?)	Ennos 1981
		VIII	predation?	Dritschilo & Pimintel 1979
		VII (including direct comparison) and VIII	predation by many mollusk species, aphids	Angseesing 1974, Angseesing & Angseesing 1973, Crawford-Sidebotham 1972, Dritschilo et al. 1979, Ellis et al. 1977, Whitman 1973, Dirzo & Harper 1982a,b
		VII	rust damage by *Uromyces trifoli*	Dirzo & Harper 1982b

NOTES:
* Data used in Chapter 7.
[p] Known to be controlled by 1–3 polymorphic loci.
[q] Known to be a quantitative genetic (continuous) trait.
[(p),(q)] Genetics strongly inferred, or known in closely related species. The remaining (unmarked) traits in this table are likely to be heritable by analogy with other species.
m: mean affected by natural selection.
v: variance affected by natural selection (in parentheses if conservative tests in Chapter 7 show that the change is not significant).

Very often a single paper on the species in question was not sufficient to justify its inclusion in Table 5.1; but taken together, several papers on the same species were includable. Therefore, in reading the original references cited for a particular species, it is advisable to consider more than one or two of the cited papers. In addition, many of the cited papers include other methods of demonstrating selection, but only the results of methods VI–VIII were used as criteria for admission. Table 5.1 yields a number of interesting patterns, which are discussed below.

5.2. OBSERVATIONS ON THE DISTRIBUTION OF KINDS OF TRAITS SELECTED

Demonstrations of natural selection are widely distributed among a great diversity of animal and plant taxa; there is no evidence that any particular group is more likely to be represented than another. More animal species are represented than plants (Table 5.2), but this probably is an artifact; until recently, studies of natural selection have been primarily done by zoologists. This is rapidly changing, and there seems to be nothing inherently more difficult about detecting selection in either kingdom. Within animals, vertebrates and invertebrates are about equally represented. Since there are more invertebrate species, this might suggest that selection is more common in vertebrates; but once again, this is probably the result of greater interest by vertebrate biologists. I hope this book will encourage correction of the inbalance.

It would be interesting to ask whether or not genera that are currently undergoing rapid radiation are more likely to show detectable natural selection than, say, species-poor genera. Unfortunately, there are insufficient data to address this question at present; it would repay further study. It would be especially interesting to study intensively natural selection of a few traits in a species-rich and a species-poor genus living in the same habitat.

Of the various components of fitness, mortality is the most commonly demonstrated, perhaps because many researchers have not looked for differences in fertility, mating ability, and other components. Yet the others are by no means rare (Table 5.1). It is impossible to say which forms of selection are more common because the number of unpublished studies with insignificant results is unknown. (This problem also applies to the other patterns discussed in this section.) However, since about 1982 there has been much interest in mate choice and sexual selection, and this aspect of fitness has only just begun to be explored seriously. Many of the examples in Table 5.1 involve mate choice, and examples are being published so rapidly that this table should be regarded as quite incomplete for sexual selection, especially after 1983. The burgeoning literature gives the impression that differences in mating success are found almost whenever they are investigated. For example, in a survey of frogs and toads, Howard and Kluge (1985) found non-random mating in 22 out of 30 species. The same may be true for fertility and fecundity effects. They have been little studied, yet O'Donald (1983) showed that fertility has a greater effect than mortality selection in determining an equilibrium. At the very least these observations suggest that reproduction-related selection may be a powerful and heretofore underestimated factor in evolution (see also Chapter 7).

Demonstrations of natural selection are equally divided among polymorphic and quantitative traits, both in terms of number of species and number of traits studied (Table 5.2).

Considering species numbers, demonstrations of natural selection are most common in morphological traits, next in physiological traits, and least common in biochemical traits (definitions and data in Table 5.2). The pattern is less uneven when the total number of traits is considered; selection in biochemical traits then becomes as common as in physiological

TABLE 5.2. Distribution of studies demonstrating natural selection (data from Table 5.1)

	Numbers of Species										
	Animals				Plants				All Species		
	Q	B	P	*T*	Q	B	P	*T*	Q	B	P
Morphological	33	4	34	*71*	4	0	10	*14*	37	4	44
Physiological	7	0	2	*9*	8	7	3	*18*	15	7	5
Biochemical	—	—	10	*10*	—	—	2	*2*	—	—	12
Two or More	0	8	1	*9*	4	4	0	*8*	4	12	1
Total	40	12	47	*99*	16	11	15	*42*	56	23	62

	Number of Traits								
	Animals			Plants			All Species		
	Q	P	*T*	Q	P	*T*	Q	P	*T*
Morphological	81	63	*144*	30	25	*55*	111	88	*199*
Physiological	18	2	*20*	22	14	*36*	40	16	*56*
Biochemical	—	43	*43*	—	16	*16*	—	59	*59*
Total	99	108	*207*	52	55	*107*	151	163	*314*

NOTES:
Q = Quantitative traits:continuously varying traits, whether or not their heritabilities are know
Mostly [q] in Table 5.1.
P = Polymorphic traits:discontinuously varying traits, whether or not their genetics are know
Mostly [p] in Table 5.1.
B = For species for which natural selection was demonstrated in both quantitative and po
morphic traits.
T = Total: Q + B + P (species) or Q + P (traits).
Morphological traits include body size, shape, etc., and behavior. Physiological traits inclu
resistance to pathogens, parasites, or herbivores; life history parameters; cyanogenesis; alkaloi
heavy metal tolerance. Some could be regarded as biochemical variants. Biochemical traits a
mostly allozymes and other electrophoretic traits.

traits. However, the relative abundance of biochemical selection demonstrations may be overestimated because some loci may be markers for selection at linked loci. For example, the allozyme loci in *Hordeum* have been shown to be selected in groups rather than individually. For both species and trait numbers, selection in physiological traits is considerably more common in plants than in animals, primarily because there is such a large component of phenotypic plasticity to plant morphology; it is comparatively difficult to find morphological traits with high heritability in plants.

Why are examples of biochemical selection comparatively uncommon? They may be uncommon because biochemical variants are usually selectively neutral, or merely because we have a difficult time assigning the selective agent and therefore identifying traits that are likely to be directly affected by selection. The distribution of heterozygosities, effective numbers of alleles, and other aspects of biochemical and molecular variation in natural populations are consistent with predictions from models assuming only random genetic drift and mutation (Nei 1983; Kimura 1983). Perhaps only selectively neutral or near-neutral variants have escaped selection to fixation or loss (but see Crow 1979); functionally less important portions of protein and DNA sequences appear to be more variable than portions that are more important in function (Nei 1983; Kimura 1983). For example, Prakash (1977) found that six populations of *Drosophila persimilis* showed more variation in nonspecific enzymes and less variation for specific enzymes such as those in the glycolytic pathway and in the Krebs cycle.

On the other hand, compared to morphological traits, we are working in a state of comparative ignorance in biochemical traits. There have been very few critical studies of the biochemical and physiological consequences of known variants, and even fewer that have yielded enough information to suggest mechanisms of natural selection (Koehn et al. 1983; McDonald 1983; Zera et al., 1985). Notable exceptions are LAP in *Mytilus edulis* mussels, PGI and α-GPDH in *Colias* butterflies, LDH in *Fundulus heteroclitus* killifish, and sickle cell and G6PD in man (references in Table 5.1). It does appear that whenever allozymes are examined closely, significant functional differences are found among alleles (McDonald 1983; Zera et al., 1985). But very few attempts have been made to relate functional differences to demonstrated natural selection; a notable exception is Watt's study of PGI in *Colias* (1977, 1983; Watt et al. 1983).

Two examples show the danger of extrapolation from functional biochemical differences in allozymes to differences in selection. (1) In a few species, genotypic differences in allozyme temperature stability in the laboratory parallels latitudinal and elevational variation in field temperatures (Koehn et al. 1983; Oakeshott et al. 1982), but there is at least one case in which the clines go in an opposite direction than predicted (Oakeshott et al. 1981). Clearly the relationship between function and fitness is complex in this case. (2) Many different cytochrome-*C*'s from different species react equally well in vitro with horse heart cytochrome oxidase, but the standard methods involve nonphysiological conditions. When the testing is repeated under physiological conditions, a given cytochrome-*C* has a marked affinity only for the cytochrome oxidase of its own species (Hartley 1979). Experimental conditions may mask field differences. In the literature on biochemical variation it is all too common to find a demonstration of natural selection conditions *a* (variation) and *c* (inheritance) without *b* (fitness differences). This is at least in part because the methods for demonstrating *b* are so different from those for *a* and *c*.

Unless we know the functions of biochemical variants (including the possibility that the functions of allelic variants are identical), we do not even know if the loci concerned are likely to be the characters directly affected by natural selection. But the relative ease of demonstrating natural selection on morphological characters, and the difficulty of demonstrating it for allozymes, might lead one to suspect that the latter are little influenced by natural selection. Perhaps the variation in morphological characters is affected primarily by "regulatory" rather than "structural" genes (King and Wilson 1975); but given the unknowns, the name "regulatory" does not explain how they work or how they evolve. As our knowledge of developmental genetics increases, perhaps natural selection will become easily detectable at the biochemical and even the molecular level. At present, the only strong evidence for natural selection at the molecular level is the spread of "selfish" DNA and the *Drosophila melanogaster* "*P*" and "*I*" factors (Doolittle and Sapienza 1980; Kidwell 1983a,b). But these behave more like successful parasites than genes favored by natural selection.

Examples of physiological factors are also uncommon compared to morphological traits. As in the case of biochemical traits, these might be so closely linked to survival and successful reproduction that there is little genotypic variation. Unfortunately, there are too few data to support or refute this conjecture. There has been very little interest by physiologists in variation within species or in the significance of that variation (discussion in Mayr 1961 and Cain 1964); therefore, natural selection on physiological variation is largely unexplored. A major exception is the work on *Mytilus* mussels, *Colias* butterflies and *Fundulus* killifish (Table 5.1).

There are four possible reasons (nonexclusive) for the relative abundance of morphological, physiological, and biochemical traits in Table 5.1: (1) Morphological traits are more frequently subject to natural selection than are physiological or biochemical traits. (2) Morphological traits, being not so rigidly involved in the internal economy, are more variable, making natural selection more easily detectable, and its results more obvious, than in the other traits. (3) We are relatively more ignorant of the function of biochemical and physiological traits, making it harder to know why and how they could be selected, than of morphological traits. Since it is more difficult to know how and why molecules may be selected, it is more difficult to design studies that will detect selection. (4) Morphological traits are examined more often for selection than biochemical traits. Reason 4 is unlikely, in view of the intensity of the selection-drift controversy in the late 1960s and 1970s. More work has to be done to minimize the effects of reason 3 and to determine the importance and truth of reasons 1 and 2.

There is a fifth possible reason. Milkman (1982) and Lynch (1984) related the effects of selection differentials of quantitative traits to the resulting selection coefficients on the loci contributing to the trait. If it is the quantitative trait that is selected, and many polymorphic loci contribute to the trait, then the effects of selection on each locus will be so small that it may be impossible to detect. Not only may selection of each contributing polymorphic locus be difficult to detect, but it may also be small enough for the loci to behave in a quasi-neutral fashion (Milkman 1982; Lynch 1984). This, together with the unknown function of most biochemical traits, might explain the relative uncommonness of direct demonstrations of natural selection of biochemical and molecular variants.

5.3. OBSERVATIONS ON THE DISTRIBUTION OF MODES OF SELECTION

Even though heterozygous advantage is the most popular theoretical means of maintaining polymorphisms (Dobzhansky 1970), I found only six examples: *Colias philodice* (butterfly), *Dendragapus obscurus* (grouse), humans, *Geomys* (pocket gophers), *Hordeum vulgare* (barley), and *Mimulus guttatus* (monkeyflower) (references in Table 5.1). It is possible that when all fitness components are considered together, there may be a *net* heterozygous advantage or stabilizing selection; but this not detectable in most studies that measure selection only at a few age classes or life-history stages. In a survey of allozyme heterozygosity in animals, Allendorf and Leary (1986) found 9 cases of increased and 6 cases of decreased survival of more heterozygous individuals with respect to one locus in 14 species. This is not significantly different from 1 : 1 expected, reinforcing the observation that heterozygous advantage is not common. Curiously, the pattern is different with respect to multiple-locus heterozygosity: they found 7 species with increased and none with decreased survival of individuals with more heterozygosity ($P < 0.05$). Milkman's (1982) and Lynch's (1984) suggestion that allozymes may contribute to selected quantitative traits might apply here. In addition, Turelli and Ginsburg (1983) suggest that most multi-locus fitness models which are capable of maintaining polymorphisms yield a correlation between fitness and heterozygosity. Unfortunately, there is no known genetic or physiological mechanism for the observation, and it would repay further study. Stabilizing selection is not particularly common either. In addition, many of the well-known studies that apparently show stabilizing selection (DiCesnola 1907; Inger 1942, 1943; Weldon 1901, 1903) suffer from a lack of control for sampling intensity among age classes (Chapter 4), and so are not included in Table 5.1. As I mentioned in Chapter 4, a reduction of variance can result from directional selection or selection of correlated traits. Finally, stabilizing selection does not maintain genetic variation (Lande 1976, 1980; Nei 1983). It is therefore likely that heterozygous advantage and stabilizing selection are neither universal nor important mechanisms that maintain genetic variation.

There are almost no examples showing competition or density-dependent selection among phenotypes. This may be because intraspecific competition is actually rare, or because there are special and very difficult problems of demonstrating competition, even between species (Harvey et al. 1983). Schoener (1983) found 148 (out of 164) successful demonstrations of competition among species in field experiments, suggesting that experimental manipulation of natural populations (method VI) is more likely to show competition than is direct observation. For further discussion of competition experiments and their special problems, see Harvey et al. (1983), Schoener (1983), and Bender et al. (1984). The apparent rarity of competitive effects within species may arise because the appropriate field experiments have simply not been done; this would repay further study.

Constant fitnesses are probably not the rule. Where data are extensive enough for it to be detectable, frequency-dependent selection is common, but such studies are uncommon. Too few researchers have planned their studies well enough to be capable of demonstrating frequency dependence; this requires extensive work with a variety of phenotype frequencies. In the few cases where selection has been measured more than once, selection is found to vary in time. There are also many examples of geographically varying selection. These observations will come as no surprise to anyone who has worked with animals or plants in the field. It is important to emphasize again the danger of assuming that selection is constant in time or space, or independent of the composition of the population. The implicit or explicit assumption of constant and geographically uniform selection is one of the last vestiges of typological thinking in population biology. On the basis of the relative frequency of occurrence shown in Table 5.1, geographically and temporally varying selection and frequency-dependent selection are more likely than heterozygous advantage to be important factors that maintain genetic variation.

5.4. GENERAL COMMENTS ON DETECTING NATURAL SELECTION

Most studies of natural selection contain three major faults: (1) no estimates of lifetime fitness; (2) consideration of only a few traits; and (3) unknown or poorly known trait function.

These are frequent symptoms of a fundamental and widespread lack of interest in the organisms, an overenthusiasm for testing one's favorite theories, and little interest in *why* natural selection can occur. This is ironic because evolutionary biologists are supposed to be more interested in *why* than in *how* questions (Mayr 1961; Cain 1964).

(1) There are few attempts to estimate lifetime fitness; the few examples are summarized by Clutton-Brock (1983). This is true at least in part because it is very difficult to study natural selection over the whole lifetime of individuals, and certain components (such as mating ability or fecundity) may be very difficult to obtain in natural populations. But this is no reason to make generalizations about the consequences of natural selection from incomplete estimates. Even the few estimates of lifetime fitness were made on traits that have no or very low heritability—for example, *Parus major* tits (McGregor et al. 1981) and *Cervus elaphus* red deer (Clutton-Brock et al. 1982). The now obvious rarity of lifetime fitness estimates for heritable traits should encourage more attempts at complete descriptions of natural selection.

(2) Few studies are concerned with more than one or two traits, and even fewer look for selective interactions among them. Three noteworthy exceptions are Clegg et al. (1978a), Lande and Arnold (1983), and Price (1984). An organism does not consist of a bag of traits, each of which can be considered in isolation. This has been stated repeatedly over the years (for example, by Robson and Richards 1936; Mayr 1942, 1963; Wright 1942, 1982) but largely ignored. Natural selection affects the whole organism, and many of an organism's traits will contribute to its ability to mate and survive. Studies considering only one or a few traits may therefore be very misleading. One reason that most studies consider few traits is that their primary interest was in demonstrating natural selection; the size of Table 5.1 shows that this is no longer a sufficient

reason. A second reason is that until recently there was no good way of analyzing selection on many traits; this has also changed (see Chapter 6).

(3) There are few cases in which it is known why natural selection occurs. As in (2), this is at least in part because most researchers were satisfied in demonstrating merely that natural selection occurred, not knowing or caring why. But this is a serious mistake. Carrying the chemical analogy (Chapter 2, section 2.2) further, this is equivalent to demonstrating a chemical reaction, and then not investigating its causes and mechanisms. A strong demonstration of natural selection combined with a lack of knowledge of its reasons and mechanisms is no better than alchemy. The most successful studies are those in which it is possible to assign a direct cause-effect relationship among an environmental factor, the organism's biology and ecology, and the trait of interest. Examples include studies of color patterns, palatability of plants to herbivores, and resistance of hosts to pests and vice-versa. In the studies where the mechanism of natural selection is known, it is not only possible to demonstrate selection, but also to predict its outcome in new localities or in experimental manipulations (examples: *Biston*, *Colias*, *Poecilia*, *Geospiza*, *Avena*, *Lotus*, and *Trifolium* in Table 5.1). In view of the ubiquity of natural selection (Table 5.1) it is now very important to know the *reasons* and *mechanisms* for natural selection. This is especially true for phenotypic selection, and most particularly for condition *b* for natural selection (fitness differences). These reasons can be obtained only by a detailed knowledge of the ecology and biology of the organisms. The time has passed for "quick and dirty" studies of natural selection.

It would be very interesting to know the following: What are the mechanisms for natural selection? What conditions favor natural selection? What kinds of traits are most likely to be selected? What is the effect of genetic interactions among traits? What is the effect of phenotypic (selective) interactions among traits? Is there any limit to the number of traits that affect fitness, and does this vary with habitat? Is there a relationship between the presence of demonstrable natural selection and genera that are currently radiating rapidly? These questions are almost completely unexplored.

5.5. SUMMARY

There are a remarkable number of direct demonstrations of natural selection in natural populations, and they are distributed among a great variety of taxa of animals and plants, and among many different kinds of traits (Table 5.1). Although sexual selection has only recently become an active area of research, there is a rapidly growing list of examples, suggesting that this form of selection may be more important than has heretofore been realized.

Heterozygous advantage is rare and stabilizing selection is not ubiquitous, in contrast to what has often been implied in the literature. Frequency-dependent selection is found in properly designed studies, but such studies, including those of density-dependent selection, are uncommon. Geographically and temporally varying selection is common. The assumption of spatially and temporally constant fitnesses is invalid and is a good example of typological thinking. Cases of selection in biochemical traits are comparatively rare, and this may be because they are selectively neutral, because they are weakly coupled to fitness, or because we do not know enough about their function to design studies that are capable of detecting biochemical selection.

The studies in Table 5.1 suggest three main problems: (1) hardly any deal with lifetime fitness; (2) very few deal with more than one or two traits; and (3) most cannot say why natural selection occurs—the reason and mechanism for condition *b* (fitness differences) is absent. These problems are symptomatic of a general and widespread lack of interest in the organisms studied and of many researchers' satisfaction with merely detecting selection. It is time to change these attitudes, and bridge the gap between population and organismal biology.

Important unanswered questions are: What are the mechanisms of natural selection? What conditions favor natural selection? What kinds of traits are most likely to be selected? What is the effect of genetic interactions among traits? What is the effect of phenotypic (selective) interactions among traits? Is there a limit to the number of traits that affect fitness, and does this vary with habitat? Is there a relationship between the presence of demonstrable natural selection and genera that are currently radiating rapidly? These questions are virtually unexplored.

CHAPTER SIX

Estimating Selection Coefficients and Differentials

Boswell: "Sir Alexander Dick tells me, that he remembers having a thousand people in a year to dine at his house; that is reckoning each person as one, each time that he dines there." Johnson: "That is about three a day." Boswell: "How your statement lessens the idea." Johnson: "That, Sir, is the good of counting. It brings every thing to a certainty, which before floated in the mind indefinitely." Boswell: "But *Omne ignotum pro magnifico est*: one is sorry to have this diminished." Johnson: "Sir you should not allow yourself to be delighted with error."

Samuel Johnson and James Boswell, 1783
(Boswell and Glover 1901, vol. 3, p. 237)

To understand natural selection, and for predictive purposes, it is not sufficient merely to demonstrate that selection occurs; we need to know its rate, at least in the populations under study. Rates are estimated and predicted from selection coefficients and differentials. Two important questions in studies of natural selection are: (1) What are the biological reasons that some trait values have higher fitness than others, and what are the biological reasons for natural selection? (2) Given that there is fitness variation and natural selection, what are the evolutionary dynamics and equilibrium configurations (if any) of the trait? Estimates of selection coefficients and differentials are required to answer these questions.

It would also be interesting to know something about the distribution of fitnesses of various traits in various organisms because of its importance in controversies about the relative importance of genetic drift and natural selection. For example, for polymorphic traits, selection overcomes the effects of genetic drift only if the selection coefficient is greater than $1/(4N_e)$ (Ewens 1979); how often is this true? The distribution of fitnesses also bears on the controversy among different neutral theories: is it more realistic to assume that loci are either entirely neutral or strongly selected, or that there are a large number of partially deleterious loci as well (summary in Nei 1983)? Finally, some criticisms of the so-called "adaptationist program" suggest that arguments about adaptation and natural selection are without supporting evidence (for example, Gould and Lewontin 1979); direct demonstrations of natural selection, along with measurements of fitness, eliminate that criticism. The purpose of this chapter is to discuss methods for estimating selection coefficients and differentials and their associated problems.

6.1. INTRODUCTION TO THE METHODS

There are many different methods for measuring selection; reviews may be found in Cook (1971), Falconer (1981), Hedrick (1983), Johnson (1976), and Wallace (1981). Most of the methods assume that fitnesses remain constant during the estimate and that the estimate is proportional to lifetime fitness. Many also assume that there is no density-dependent selection and that generations are nonoverlapping. As mentioned in Chapters 2 and 4, these assumptions can lead to serious problems. All methods are only as valid as their assumptions.

For all methods it is important to make the distinction between absolute and relative fitness, and this applies to both polymorphic (Ewens 1979; Wallace 1981) and quantitative (O'Donald 1970, 1971; Lande 1979; Lande and Arnold 1983) traits. Let the absolute fitness of phenotype (or genotype) X be $W(X)$, and its relative fitness be $w(X)$. Relative fitness is $w(X) = W(X)/\bar{W}$, where $\bar{W} = [\sum f(X)W(X)]/[\sum f(X)]$ is the mean absolute fitness, and $f(X)$ is the frequency of genotype or phenotype X. Therefore, mean relative fitness $\bar{w} = 1$. Relative fitness can also be measured with reference to a particular phenotype (or genotype), in which case $\bar{w}$ is not necessarily 1; this is the most common method used for polymorphic traits. If the population is sampled twice (or more) within a generation so that the individuals in the second sample represent a subset of those sampled in the first sample (as in a capture-recapture or cohort study), then absolute fitnesses can be calculated. Examples are the probability of surviving between samples, or the probability of mating. On the other hand, if samples are made without replacement, or if samples are made of juveniles and adults at a single time, then only relative fitnesses can be calculated; information on total numbers and mean fitness is lost (see discussions in O'Donald 1971, Hiorns and Harrison 1970, and Manley 1974).

6.1.1. Polymorphic Traits

There are many methods for estimating fitness (w) or selection coefficients ($s = 1 - w$) of polymorphic traits. These are well known and so will not be summarized here.[1] The polymorphic trait methods fall into two major categories: (1) those that utilize changes in genotype (or phenotype) numbers or frequencies directly, and (2) those that estimate fitness on the basis of changes in gene frequency.

The direct methods can be used in both cohort and age-class methods (VII and VIII). Perhaps the most efficient method is fitness component analysis in which estimates are made between a series of adjacent life-history stages, yielding most or all components of fitness (references in Chapter 3, methods VII and VIII; summary in Hedrick 1983). A particularly clear example is found in Clegg et al. (1978a).

The methods in the second category are most appropriate for long-term study (method V) and especially for perturba-

[1] Summaries of a variety of methods are found in Cook 1971; Hedrick 1983; Hiorns and Harrison 1970; Johnson 1976; Kempthorne and Pollack 1970; Manley 1974; Prout 1965–1971a,b; Wallace 1981.

tion methods (vi) since they depend upon changes in frequency. If the genetics and ecology of the species are sufficiently well known, then an observed change in allele frequency over many generations can be fitted to a model of selection. Provided the range of gene frequency change is large enough, relatively few different selection models or sets of fitnesses will explain the change, and this may be regarded as a good estimate of fitness. A frequent problem arises if there is only a small change in frequency; a large number of different models can produce the same small change. (See section 3.5.2 for further discussion). Of course this method is only as good as the similarity between the model and nature. It has not been used in field studies as often as it might.

6.1.2. Quantitative Traits

Since quantitative traits are continuously distributed (by definition), fitnesses are rarely estimated for each trait value or trait value class, though this is a pity. It is common, instead, to estimate rates of phenotypic change. There are many measures of selection for quantitative traits, depending upon whether one wishes to consider changes in trait value mean, the variance, or both mean and variance. All of the measures depend upon changes in phenotypic frequencies within a generation and were originally developed for patterns of mortality. They can, however, be modified to include the effects of variation in fertility, fecundity, and mating ability. Since these are relatively little known outside quantitative genetics, they will be reviewed in some detail. The reader who is not interested in the details of the methods should skip to sections 6.4.4 or 6.5, where the problems and uses of the methods are discussed.

6.1.3. Quantitative Trait Model and Symbolism

Consider a trait that varies in value X, with a frequency distribution before selection of $f(X)$, with mean $\bar{X}_b$, variance

v_b, and sample size N_b. After selection, the distribution has a new mean $\bar{X}_a$, variance v_a, and size N_a ($n = N_b + N_a$). Modification for differential fecundity, fertility, or mating ability is simple. For example, if a certain range of X values has a mating advantage, the "before selection" $f(X)$ is the distribution of phenotypes without regard to their mating ability, and the distribution "after selection" is determined by who actually mated. Three groups of methods will be discussed: direct univariate, univariate mean fitness, and multivariate methods.

6.2. DIRECT UNIVARIATE METHODS

6.2.1. Standardized Selection Differentials

These methods are the simplest and were developed by animal breeders. Two measures of selection are:

$$i = \frac{\bar{X}_a - \bar{X}_b}{\sqrt{v_b}} \tag{6.1}$$

$$j = \frac{v_a - v_b}{v_b}. \tag{6.2}$$

The quantity i is a measure of directional selection, and j is a measure of stabilizing (if negative) or disruptive (if positive) selection. The quantities i and j measure the proportional change in the mean and variance, respectively. The measure i is also known as the standardized selection differential (Falconer 1981). For brevity, henceforth, stabilizing and disruptive selection will be collectively called variance selection.

If X is normally distributed, then the statistical significance of directional and variance selection can be tested. For i:

$$t_{(n-2)} = \frac{\bar{X}_a - \bar{X}_b}{\sqrt{\dfrac{n[(N_b - 1)v_b + (N_a - 1)v_a]}{(n-2)N_b N_a}}}, \tag{6.3}$$

and for j:

$$F_{\{N_b,N_a\}} = \frac{v_b}{v_a} \text{ or } F_{\{N_a,N_b\}} = \frac{v_a}{v_b} \quad \text{(whichever is larger).} \qquad (6.4)$$

These tests (two-tailed) will reveal if there is any significant directional, stabilizing, or disruptive selection. The t or F values can also be used to estimate confidence intervals for i and j (Sokal and Rohlf 1981).

The F test is extremely sensitive to departures from normality. Van Valen (1978) suggested three alternate tests, and, of those, Smith's test is explicitly designed to test the null hypothesis of equality of variances. For each group g of size N_g, calculate E_g, the variance of the estimate of v_g:

$$E_g = \frac{\sum_{i=1}^{N_g} (X_{ig} - \bar{X}_g)^4 - v_g^2(N_g - 3)/N_g}{(N_g - 2)(N_g - 3)}; \qquad (6.5a)$$

then the expected value of

$$x^2 = \sum_g \frac{v_g^2}{E_g} - \frac{\left[\sum_g (v_g/E_g)\right]^2}{\sum_g (1/E_g)} \qquad (6.5b)$$

is distributed as χ^2 with one degree of freedom. For more serious departures from normality, nonparametric statistics should be used (Siegel 1956).

In the form presented, the tests for the significance of i and j are conservative because they compare distributions before and after selection. The individuals after selection ($g = a$) are a subset of the population before selection ($g = b = a + d$). A better and more sensitive test can be made by replacing group $b = (a + d)$ with the unselected individuals (d) and comparing them to the selected group (a). The means and variances of samples $a + d$ and a will be more similar to each other than between a and d, so the $a + d$ versus a test will not always detect selection when it has occurred. The a versus d test is also superior because the two samples are independent, which is formally required by the assumptions of equations 6.3–6.5. However, this comparison is possible only in special cases such as studies of cohorts, or in experiments in which the total numbers of live and dead (or mated and unmated) are known after the interval of selection. In many cases this is not possible, and the less sensitive $a + d$ versus a comparison will have to be made.

6.2.2. *Selection Coefficients or "Gradients"*

Because directional selection leads to a reduction in variance (Falconer 1981), Lande and Arnold (1983) suggest using multiple regression to allow independent estimates of selection on the mean and variance. If mean fitness is not known, this suggests an approximate correction for j:

$$j' = j + i^2. \tag{6.6}$$

The quantity i^2 is proportional to the amount that the variance is reduced by directional selection. It is a valid correction if the trait values are normally distributed and is a useful approximation for other distributions (Lande and Arnold 1983). If the variance in fitness or the mean fitness is known, and the trait is normally distributed, then the complete formulae are

$$i^* = \frac{i}{\sqrt{P}} \tag{6.7a}$$

$$j^* = \frac{j + i^2}{\sqrt{P}}, \tag{6.7b}$$

where P is the variance in relative fitness w. Equations 6.7 are the standardized multiple regression coefficients of relative fitness on X and $(X - \bar{X})^2$, with a constant factor $1/\sqrt{2}$ taken out of j^* (see Appendix 1 for derivation). If X is not normally distributed, then the formulae are more complex (Appendix 1, equations A.15 and A.6). The unstandardized multiple regression coefficients (Appendix 1, equations A.15, A.20) were called "selection gradients" by Lande and Arnold (1983). Since this term can easily be confused with selection gradients in the sense of geographic gradients in selection (the older meaning, Endler 1973, 1977), these "gradients" will henceforth be called "selection coefficients," or simply regression coefficients.

If there are only two absolute fitness classes ($W = 0$ or $W = 1$), then $P = [(1 - \bar{W})/\bar{W}][N_b/(N_b - 1)]$, where $\bar{W}$ is the mean absolute fitness, and also $\bar{W} = N_a/N_b$ (see Appendix 1). P is also known as the "potential" or "opportunity" for selection (see section 6.3.3), so it is intuitively satisfying that P is used with the standardized selection differentials (i and j) to convert them to selection coefficients (i^* and j^*). If the variance of relative fitness is standardized to unity, then equations 6.7a and 6.7b are equivalent to equations 6.1 and 6.6. The standardized selection differentials are therefore selection coefficients in terms of unit variance of fitness.

Like selection differentials, selection coefficients can also be given standard errors and tested for significance. This is much better than using equations 6.3 and 6.5 because it removes the effects of directional selection on j. Formulae for the standard errors of i^* and j^*, and details of the statistical tests, are given in Appendix 1 (equations A.9–A.11). These tests should be performed in preference to equations 6.3 and 6.5 whenever possible.

6.2.4. Useful Properties of Selection Differentials and Coefficients

Selection differentials and coefficients have the advantage over other kinds of selection measures (see the next section) in that they can be used with the heritability of the trait to yield the genetic response to directional selection, rather than merely the phenotypic selection within a generation. This yields the rate of selection. If i applies to the parents, h^2 is the heritability of the trait, and R is the deviation of the mean value of the offspring of the selected parents from the population mean, then the standardized response is

$$\frac{R}{\sqrt{v}} = ih^2 \tag{6.8}$$

(Falconer 1981), provided that selection does not affect any traits that are genetically or phenotypically correlated with the trait of interest. If there is covariance with other traits, then the predicted change is:

$$\frac{R}{\sqrt{v_1}} = b'_1 V_1 + \sum b'_k C_{k1}, \tag{6.9}$$

where the b'_k are the standardized multiple regression coefficients of relative fitness on trait value X_k, and C_{k1} is the genetic covariance between the trait of interest (trait 1 with genetic variance V_1) and trait k. See Lande (1979), Lande and Arnold (1983), and section 6.4 for a discussion of phenotypic selection on phenotypically correlated traits; see Bulmer (1980) for a discussion of the effects of genetic correlations on the genetic response. Phenotypic or genetic correlation with other traits may yield i and R values that are larger or smaller than the actual value. This problem applies to all other single-trait methods as well.

Kimura and Crow (1978), Crow and Kimura (1979), Milkman (1982) and Lynch (1984) discuss the relationship between selection on polymorphic and quantitative traits. If it is possible to order the genotypes at a single locus on the value axis, then there is a simple approximate relationship $s_k = ig_{kl}$, where g_{kl} is the phenotypic effect of substituting genotype k for the reference genotype l ($s_l = 1.0$). Here, g_{kl} is the difference between the mean effects of genotypes k and l on the quantitative trait value X (Milkman 1982). If a logical ordering of the genotypes is not possible (g undefined, as would be true for color pattern genotypes, for example), or there are significant nonadditive effects, then the relationship does not work, though it is quite reasonable to use it for many loci, including allozymes.

6.3. UNIVARIATE MEAN FITNESS METHODS

These methods require estimates of fitness or assumptions about fitness $W(X)$ for each value X (or each X class), and a single statistic estimates the effects of *both* directional and variance selection. There are two approaches: (1) percentage of deaths due to individuals not being at an optimum trait value, and (2) change of mean fitness due to selection on the trait.

6.3.1. Suboptimum Deaths

This method assumes that there is some optimum value m at which fitness is maximized, $W(m)$. A measure of selection is the minimum reduction of the original population that would yield a population with specified values of the parameter or set of parameters (Haldane 1954; Van Valen 1963a). It was first proposed by Haldane (1954), who defined the measure as

$$H = \ln[W(m)/\bar{W}], \tag{6.10}$$

and later improved by Van Valen (1963a, 1965a) as

$$I = \frac{W(m) - \bar{W}}{W(m)}. \tag{6.11}$$

Equation 6.11 has the more obvious meaning: The percentage of deaths due to not being at the optimum value m.

Haldane's measure is related to Van Valen's by

$$I = 1 - \exp(-H), \tag{6.12}$$

when $W(m) = 1$, and both measures will be similar for small I (Haldane 1954; Van Valen 1965a; O'Donald 1970).

If the distribution of X values is normal both before and after selection, Haldane (1954) showed that

$$H = \frac{1}{2}\ln\left(\frac{v_b}{v_a}\right) + \frac{(\bar{X}_a - \bar{X}_b)^2}{2(v_b - v_a)}. \tag{6.13}$$

To avoid the necessity for this assumption, Van Valen (1965a, 1967) presented charts for estimating I for any combination of $\bar{X}_b$, $\bar{X}_a$, v_b, v_a, but with the assumption of truncation selection—that is, his charts give the selection rate that would yield the observed distribution after selection if truncation selection occurred. A problem with this method is that selection would almost never operate on absolute thresholds in natural populations. If the actual selection mode, or fitness function $W(X)$, does not show a threshold, then I will *underestimate* the rate because fewer deaths are needed in truncation compared to nontruncation selection to cause the same shift in mean and variance (O'Donald 1968, 1970).

O'Donald (1968, 1970, 1971), Cavalli-Sforza and Bodmer (1971), and Manley (1975–1977) avoided the assumption of truncation selection by estimating I in conjunction with an explicit assumption about the relationship between the value X of a trait and its fitness, $W(X)$:

Linear model (O'Donald 1970):

$$W(X) = J + KX; \tag{6.14}$$

Double exponential (Manley 1976):

$$W(X) = \exp[\exp(J + KX)]; \tag{6.15}$$

Quadratic (O'Donald 1968, 1970, 1971):

$$W(X) = 1 - a - K(m - X)^2; \tag{6.16}$$

Normal (Cavalli-Sforza and Bodmer 1971):

$$W(X) = (1 - a)\exp[-K(m - X)^2]. \tag{6.17}$$

Here J, K, and a are constants. The quantity a is 1 − absolute fitness at the optimum ($X = m$), and may be set to zero if calculations are to be performed on relative fitnesses (O'Donald 1970). In the quadratic and normal models, fitness reaches a maximum at an intermediate trait value m, and falls off for larger and smaller values, while in the linear and double exponential models fitness either increases or decreases with X. Thus, the quadratic and normal models are most appropriate for systems in which stabilizing or disruptive selection may occur, whereas the other two are most appropriate for situations in which directional selection predominates, or if the optimum value is outside the range of X in the population studied. In all cases, I estimates the combined effects of directional and variance selection. Because there is no optimum value in the linear and double exponential models, they cannot be used to estimate I, although they are useful for estimating the change in fitness of a population (see below). Manley (1975) also discusses more complex models that include the effect of one genotype on another (competition); all other models assume that the fitness of an individual with a value X is not affected by the presence of other genotypes of any value. Estimates of I (equation 6.11 and other methods) are described with worked examples for the various models by O'Donald (1968, 1970, 1971) and Manley (1975–1977).

6.3.2. *Change of Mean Fitness*

If $f(X)$ and $W(X)$ are known before and after selection, then $\bar{W}_b$ and $\bar{W}_a$ can be calculated directly for two successive samples. A measure of the relative change in $\bar{W}$ as a result of selection during the period between the samples is

$$I_w = \frac{\Delta \bar{W}}{\bar{W}_b} = \frac{\bar{W}_a - \bar{W}_b}{\bar{W}_b} = \frac{v_b}{\bar{W}_b^2} \tag{6.18}$$

(O'Donald 1970). This is another statement of Fisher's (1930) fundamental theorem of natural selection, leaving out the effects of heritability (Crow 1958; O'Donald 1970). If $f(X)$ and $W(X)$ are not known before and after selection, then I_w can be estimated assuming one of the models (equations 6.14–6.17).

For the *linear model* (equation 6.14),

$$I_w = \frac{(\Delta \bar{X})^2}{v_1} = i^2 \tag{6.19}$$

(O'Donald 1970). The relationship to i comes from equation 6.1. Thus this is one of the few measures that can be used for prediction of the response to phenotypic selection. The relationships between I_w and i or j are not so simple in the other models.

6.3.3. *The Potential or Opportunity for Natural Selection*

Instead of measuring the observed rate of selection, we can estimate the maximum rate that is possible in a given population, given its genetic variation, population structure, and so forth (Crow 1958). Some of these measures of selection are identical to and others are related to I and I_w. For example, P (equation 6.3 and Appendix 1, equation A.22) is the variance in relative fitness, one measure of the potential for selection (Crow 1958; Arnold and Wade 1984a). If there are only two fitness classes, then this is parametrically equal to $(1 - \bar{W})/\bar{W}$ (see Appendix 1), one version of I. However, the potential for selection is more concerned with predicting future response to selection than what is happening or has happened to a population. For example, in the linear model (equation 6.14) $|i| \leq \sqrt{P}$ (Arnold and Wade 1984a); P only says what the maximum possible i can be. A detailed review, with numerous examples in humans, is found in Spuhler (1976).

6.4. MULTIVARIATE METHODS

Selection does not occur on single traits in isolation; many traits together affect the success or failure of an individual. In addition, traits may be phenotypically and genetically correlated with each other. Selection on one trait may yield apparent phenotypic selection on a phenotypically correlated trait (Chapter 3, Figure 3.2), and will yield a correlated response to selection if there is also a genetic correlation among the traits (Bulmer 1980; Falconer 1981). Correlated response to selection can be investigated by genetic analysis (Hegmann and Dingle 1982 is an example in a natural population). Effects of phenotypic correlations in phenotypic selection can be accounted for by two related multivariate methods, discriminant function analysis and multiple regression.

6.4.1. Discriminant Function Analysis

Discriminant function analysis finds linear combinations of the simple X variables which best divide the population into known groups (Lachenbruch 1975). For analysis of natural selection, there will often be only two groups: those that survived (or mated) and those that died (or did not mate). The discriminant function is an estimate of the form or mode of selection, since it yields a line or surface separating individuals of high and low fitness. For continuously varying fitness (e.g. fecundity) the population could be divided into two (split at the mean or median), and the discriminant function would then give an isoline for the fitness surface.

There are a number of different ways of calculating the discriminant function, depending upon how much is known about the distribution of values (for an especially clear review, see Karson 1982). For studies of natural selection, we do not know the means, variances, and covariances among the traits; they are estimated from the data. This requires a function different from the one often calculated in some of the commercially available "statistical packages," and is performed as follows: (1) Let N_a be the number of individuals that survived (or mated), and N_d be the number that did not; $N = N_a + N_d$ (note the difference in N and the definition of the groups from those in equations 6.1–6.19; in the latter, b and a refer to the same population before and after selection; here a and d refer to the favored and unfavored groups). (2) Let $\mathbf{x}$ be the column vector (set) of X values for all traits, measured on a single individual. (3) Calculate $\bar{\mathbf{x}}_g$, the vector of mean values for each trait of individuals in group g ($g = a, d$). (4) Calculate $\mathbf{A}_g$, a matrix containing the "corrected" sum of squares and cross-products for group g. Its elements are $a_{kl} = \sum X_k X_l - \sum X_k \sum X_l / N_g$

for traits k and l; for example, a_{kk} is $(N_g - 1)$ times the variance of trait k. (5) Calculate the estimated joint (phenotypic) variance-covariance matrix $\mathbf{V} = (\mathbf{A}_a + \mathbf{A}_d)/(N - 2)$; each element is $v_{kl} = (a_{akl} + a_{dkl})/(N - 2)$. (6) The discriminant function coefficients are the elements of the vector:

$$\mathbf{d} = \mathbf{V}^{-1}(\bar{\mathbf{x}}_a - \bar{\mathbf{x}}_d), \tag{6.20}$$

where $\mathbf{V}^{-1}$ is the inverse of $\mathbf{V}$, and $(\bar{\mathbf{x}}_a - \bar{\mathbf{x}}_d)$ is a vector of differences between the means of the groups for each trait. The constant in the discriminant function equation is:

$$C = (\bar{\mathbf{x}}_d' \mathbf{V}^{-1} \bar{\mathbf{x}}_d - \bar{\mathbf{x}}_a' \mathbf{V}^{-1} \bar{\mathbf{x}}_a)/2 - \ln (N_d/N_a). \tag{6.21}$$

The discriminant function is therefore

$$D = C + d_1 X_1 + d_2 X_2 + \cdots + d_t X_t \tag{6.22}$$

for t traits. Each individual will have an associated value of D; if $D > 0$, then the discriminant function assigns the individual to group a, and if $D < 0$ it is assigned to group d, in such a way as to minimize classification errors (Lachenbruch 1975; Karson 1982). D defines a hyperplane separating live and dead (or mated and unmated) individuals, and therefore estimates the geometric form of selection on many traits. Note that in this formulation the mean value of D for all N individuals is not zero; it is shifted so that the separation point is at zero. Many computer "statistical packages" give other forms of the discriminant function, for example, "corrected" d_k, which are sometimes $d_k\sqrt{v_k}$ and sometimes functions of $(X - \bar{X}_k)/\sqrt{v_k}$ rather than X.

If there are only two traits, then there are only two dimensions, and D defines a line. Letting $t = 2$, setting equation 6.22 equal to zero, and solving for X_2, the line separating live and dead groups is

$$X_2 = \frac{-C}{d_2} - \frac{d_1}{d_2} X_1. \tag{6.23}$$

If selection occurs with respect to an absolute threshold, then D is an estimate of the position of the threshold. If selection occurs with respect to "gradient" (gradual function of the trait values), then D estimates the position of the midpoint of the gradient. The discriminant function can be drawn perpendicular to this line, going through the mean of both groups $(\bar{X}_1, \bar{X}_2)$.

In reanalyzing Bumpus's (1899) data on selection on some morphological traits in house sparrows, O'Donald (1973) used the discriminant function axis as an X variable for estimating I and I_w, in addition to his analysis of the individual trait values. The discriminant function yielded larger estimates than did any of the individual traits. The relative magnitude of I, i, or j measures and their values on the D scale, will depend upon the scaling of D and the degree of correlation among the traits.

6.4.2. Multiple Regression Analysis

A more direct estimate of the contribution to fitness of many traits can be obtained by multiple regression. Cornfield (1962), working on "heart attack" risk; Manley (1975–1977), building on earlier work by Cox (1972), and Lande and Arnold (1983), Arnold (1983a); Arnold and Wade (1984a,b); and Arnold et al. (1986), building on some earlier work by Pearson (1903), independently devised models and methods to take covariance among traits explicitly into account. In each, an estimate is made of the function relating the fitness of an individual (W) to the values X_k of each trait k:

$$\text{Lande-Arnold model: } W = a + \sum b_k X_k; \tag{6.24}$$

$$\text{Cornfield model: } W = 1/[1 + \exp(a + \sum b_k X_k)]; \tag{6.25}$$

$$\text{Manley model: } W = \exp[-f(t) \exp(a + \sum b_k X_k)], \tag{6.26}$$

where a and the coefficients b_k are constants relating W to each X_k, holding the other X_k constant, and $f(t)$ is a positive, nondecreasing function of time t such that $f(0) = 0$. This $f(t)$ was chosen because Manley's model was explicitly designed for only the survival component of natural selection (W = the probability of surviving to time t), which cannot increase with time. The double exponential function was chosen for the same reason, and also because Manley designed the method for repeated sampling of cohorts: if p_1 is the probability of surviving through one time unit, then $p_1^{f(t)}$ is the probability of surviving through time t, and it is desired that $0 \leq W \leq 1$. Cornfield chose the logistic form because of the discontinuous nature of the classification into high and low fitness (see also Miller et al. 1981). Van Valen (1963a) and Marcus (1964) considered multivariate modifications of H and I for multiple traits; these are geometrically related to the other multivariate methods. Unlike the other methods, the Lande-Arnold model is general and does not require an explicit assumption about the relationship between W and X, though (6.24) will only estimate the linear components of selection.

The Lande-Arnold model works for all components of fitness (Lande and Arnold 1983; Arnold 1983a; Arnold and Wade 1984a,b), and the other models can be modified for fitness components besides survival. Lande and Arnold (1983) and Cornfield (1962) estimate a and the coefficients b_i by means of multiple regression (least squares), while Manley (1976, 1977) estimates them by a maximum-likelihood method. Cornfield's (1962) method is now common in the medical literature on disease risk, and Manley's method is one of a wide variety of methods in the materials-risk analysis literature (see Miller et al. 1981). Manley (1977) utilized the variance of W among individuals as an estimate of total selection, in addition to examining the individual regression coefficients. Price (1984) performed an analysis using the logit transformation, and Cornfield (1962) used discriminant functions as well as the logistic transformation. Other transformations, maximum likelihood methods, and nonparametric methods are summarized in the advanced text by Miller et al. (1981).

To elaborate on Lande and Arnold's (1983) method: A multiple regression is made using relative fitness as the dependent variable (w), with each trait value as the independent (predictor) variable, or

$$w = a + b_1 X_1 + b_2 X_2 + \cdots + b_t X_t \tag{6.27}$$

for t traits. As in the univariate case (Appendix 1), absolute fitnesses (W) are divided by $\bar{W}$ to yield relative fitnesses w with $\bar{w} = 1$. For example, if the measure of fitness is survival or not, then W has a value of 1 or 0, respectively, and w is W divided by the overall proportion surviving ($\bar{W}$). W can also be the number of days survived, the fecundity, relative mating success, or other measure of fitness (for examples, see Arnold 1983a; Arnold and Wade 1984b). The b_k coefficients ("gradients") estimate the contribution to fitness of trait k, holding the effects of other traits constant. Standardized coefficients, b'_k (see Appendix 1, equation A.6), allow comparison of degree of importance among traits and species (Sokal and Rohlf 1981; Karson 1982) as well as prediction of the genetic response (Lande and Arnold 1983). (Note that Lande and Arnold 1983 standardized their X_k variables before analysis; this yields b'_k directly, but at the expense of greater rounding errors.) For this analysis to be valid, the usual assumptions of regression must be true; in particular, the errors around the regression surface must be uniform over all values of the X's (homoscedastic), and the X's must be multivariate normal (Lande and Arnold 1983; Sokal and Rohlf 1981). Appendix 1 (equations A.16, A.17) shows one effect of nonnormality in an example with two traits. The regression coefficients are estimates of contributions of each trait to overall fitness, and can be tested for significance in the usual way (Appendix 1, equations A.9–A.11). The coefficients in both the Lande-Arnold and Manley methods estimate directional selection components.

In addition, Lande and Arnold (1983) show that quadratic multiple regression can also be used to estimate the stabilizing or disruptive components of selection on each trait, after removing the effects of directional components. This is similar to O'Donald's method for single traits (equation 6.16), though he does not have a separate term for each component. A quadratic multiple regression is done by adding X^2 or $(X_i - \bar{X})^2$ terms as extra variables (see Appendix 1, sections 2 and 5). The reason that this works is most easily seen by considering the geometry of multivariate selection as shown by the discriminant function (see the next section). In a similar way, correlational selection between traits i and j can be estimated by adding X_iX_j terms to the multiple regression, $X_iX_jX_k$ for traits i, j, and k and so on (Arnold et al. 1986).

6.4.3. *Relationship between Discriminant Functions and Multiple Regression*

When observed fitnesses have two values (survived or not, mated or not), the multiple regression of observed fitness on trait values is algebraically related to the discriminant function separating the two groups. (This is only true when there are two groups; see Lachenbruch 1975.) The relationship between d_k (the kth discriminant function coefficient, in equation 6.22), and b_k (the kth unstandardized multiple regression coefficient, in equation 6.27) depends upon the Mahalanobis distance (D^2) between the two groups:

$$D^2 = (\bar{\mathbf{x}}_a - \bar{\mathbf{x}}_d)'\mathbf{V}^{-1}(\bar{\mathbf{x}}_a - \bar{\mathbf{x}}_d), \quad (6.28)$$

and is

$$b_k = \frac{P'}{1 + P'D^2} d_k, \quad (6.29)$$

where $P' = \{\sum w^2 - (\sum w)^2/N\}/(N - 2)$, the corrected variance of relative fitness (from Karson 1982). Thus (when there are two groups) the multiple regression and discriminant function coefficients are proportional.

If survivors are assigned an absolute fitness (W) of 1 and nonsurvivors are assigned a fitness of 0, then P' simplifies to $[(1 - \bar{W})/\bar{W}] \cdot [N/(N - 2)]$. Note that for calculating b'_k it does not matter whether the survivors' $W = 1$ or some other value, since the b'_k are standardized by P'.

The Mahalanobis distance can also be measured between group b (the population before selection) and group a (after selection), rather than between groups a and d (favored and unfavored in equation 6.28). When used with a table of the truncated normal distribution, the a–b distance yields a multivariate version of Van Valen's I_w (Marcus 1964). Schluter (1984) generalized this concept, and used this Mahalanobis distance to estimate the selection required to shift the trait distribution between that observed in two different populations or species, using the dubious assumption of a constant variance-covariance matrix. This assumption is inadvisable because the form of the variance-covariance matrix can change if the selective environment changes, as was demonstrated experimentally in *Drosophila* by Service and Rose (1985). Elements of the matrix may change even under constant selection as genetic variance is reduced; the constant matrix assumption may only be valid for very weak and nearly constant selection. More work needs to be done to explore these problems theoretically and experimentally.

Quadratic multiple regression also yields a discriminant function that can separate the fit and unfit, in this case on the basis of differences in variances as well as in means. An example is shown in Figure 6.1, a case of "pure" stabilizing selection of a single trait. It is clear that the success of separation will depend upon the X^2 transformation, and in some cases other transformations may work better. However, as Lande and Arnold

(1983) point out for multiple regression, the addition of quadratic terms to the variable list is completely general and not dependent on any particular selection model. The quadratic terms explicitly estimate changes in the variance because the variance is a quadratic function of X (see also Appendix 1, section 5) and because the average of the term $(X - \bar{X})^2$ is the variance by definition.

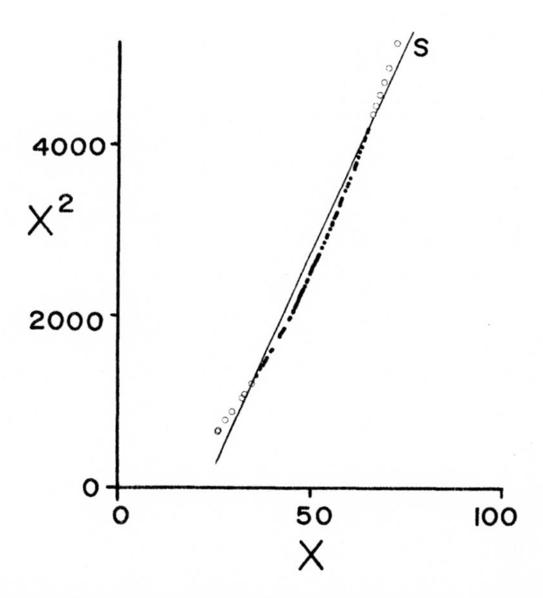

FIGURE 6.1. Discriminant function homologue of quadratic regression. The X values of the open and solid dots were generated at random from a normal distribution with mean of 50 and standard deviation of 10. The plot is X versus X^2. Solid dots are within mean ± 1.5 standard deviations on the X scale. The distribution is therefore similar to one of (1) stabilizing selection if the open dots are the individuals that die (or do not mate), or (2) disruptive selection if the solid dots are those that die. If the figure is regarded as stabilizing selection and survivors have a fitness of 1 and others of 0, then the multiple regression of fitness on X and X^2 is $w = 0.2153X - 0.0021X^2 - 4.3277$, with both b coefficients significant. The discriminant function yields a separation line $X^2 = 100.55X - 2275.81$, which is plotted and marked S. It clearly separates live and dead.

There is another direct approach to estimating selection of the trait variances, quadratic discriminant functions (Lachenbruch 1975). The ordinary or linear discriminant functions assume that the variance-covariance matrices of both groups are equal. However, if selection affects the trait variances, then this assumption is violated. The discriminant function then becomes

$$\begin{aligned}(N_d - 1)\mathbf{x}'\mathbf{A}_d^{-1}(\mathbf{x} - 2\bar{\mathbf{x}}_d) - (N_a - 1)\mathbf{x}'\mathbf{A}_a^{-1}(\mathbf{x} - 2\bar{\mathbf{x}}_a)\\ + (N_d - 1)\bar{\mathbf{x}}'\mathbf{A}_d^{-1}\bar{\mathbf{x}}_d - (N_a - 1)\bar{\mathbf{x}}'\mathbf{A}_a^{-1}\bar{\mathbf{x}}_a\\ -2\ln\left(\frac{N_d\sqrt{|\mathbf{V}_a|}}{N_a\sqrt{|\mathbf{V}_d|}}\right),\end{aligned} \tag{6.30}$$

where $|\mathbf{V}_a|$ and $|\mathbf{V}_d|$ are the determinants of the variance-covariance matrices of groups a and d (Karson 1982). This is a nuisance to calculate, and yields separation curves that are quadratic, hyperbolic, or even elliptical functions of the trait values. For example, for two traits, the separation lines may be of the form $b_1X_1^2 + b_2X_1 + b_3X_1X_2 + b_4X_2 + b_5X_2^2 + C = 0$. This is related to the "saturated" multiple regression equation, which includes the cross-products (X_kX_l), the terms that estimate correlational selection. Equation 6.30 is also geometrically related to the multivariate version of Haldane's H (Marcus 1964), if the function is calculated between the trait distributions before and after selection (b and a) rather than the unselected and selected (a and d).

The lines or surfaces from equation 6.30 will often give better separation than the linear discriminant function. However, if the differences between the means are large, or $\mathbf{V}_a$ is close to $\mathbf{V}_d$, then the linear discriminant function will perform well. The quadratic discriminant function performs quite poorly with small sample size. For further discussion of the properties of linear and quadratic discriminant functions, see Lachenbruch (1975), Lachenbruch and Goldstein (1979), and Karson (1982). A useful illustration is found in Van de Geer (1971, pp. 263–266).

Although multiple regression and discriminant function methods were developed for quantitative (continuous) traits, they can also be used for polymorphic traits, where the intercorrelations are now coefficients of gametic phase (linkage) disequilibrium. There are also some special methods in both discriminant function analysis and multiple regression that are designed to work on discontinuous X variables. Nonparametric methods may also be used for estimating the contribution to fitness of various polymorphic loci, since the genotype values will not be normally distributed.[2]

6.4.4. *Advantages of Multivariate Methods*

Multivariate methods have the great advantage in that they entirely avoid the problems that result from estimating selection on individual traits separately, and then combining them in some arbitrary way (problems discussed in Wallace 1981, Milkman 1982). For example, the problem of "genetic load" is entirely avoided by this analysis. During the "genetic load" controversy (see Wallace 1981), fitnesses with respect to different loci were combined in a completely arbitrary way. This is no longer necessary because, given a particular distribution of alleles at many loci, multiple regression estimates show how a particular combination of trait values affects the probability of survival (or some other component of fitness.) The combinations can be explicitly estimated.

The multiple regression method of Lande and Arnold (1983) has the additional advantage of an explicit relationship between b'_k coefficients and the predicted response to selection of the traits.

$$\Delta \mathbf{x} = \mathbf{G}\mathbf{B}, \tag{6.31}$$

where **G** is the genetic variance-covariance matrix, and **B** is the vector of b'_k coefficients. This is a multivariate generalized

[2] For discussions of both discrete and nonparametric methods, see Lachenbruch 1975; Karson 1982; Lachenbruch and Goldstein 1979; Goldstein and Dillon 1978; Miller et al. 1981.

version of equations 6.8 and 6.9 (see Lande 1979 and Lande and Arnold 1983 for details).

The multivariate methods and the existence of correlations among traits suggest that we must treat univariate estimates of selection with caution. A selection measure of one trait may actually reflect selection for other correlated traits, and the actual contribution to fitness of the first trait may be very different, or even in the reverse direction; this was actually found in data from *Euschistus variolarius* bugs (Lande and Arnold 1983). The b'_k coefficients indicate the contribution to fitness by that trait, holding the other contributions constant, while i^* and j^* estimate both the direct and indirect components of selection on that trait. Thus the selection coefficients (b'_k) are better estimates of the direction and rate of selection than the selection differentials (i and j), or the univariate selection coefficients (i^* and j^*) although i may be useful for estimating the response to selection if h^2 is high.

6.4.5. Problems of the Multivariate Methods

Both Manley's and Lande and Arnold's multiple regression methods were applied to the much-analyzed Bumpus (1899) data on *Passer domesticus* sparrows (Table 5.1), with similar results, suggesting, as do Lande and Arnold (1983), that the methods are fairly robust. Lande and Arnold found that only body weight and total length were significant, while Manley (1976), who did not include weight in his analysis, found total length and humerus length coefficients to be significant. The differences in results can be attributed to (1) the differences between the assumed fitness functions; (2) using or not using the logarithms of the raw data; (3) the presence or absence of a major variable (weight); and (4) the difference between the least-squares and maximum-likelihood methods of estimating the coefficients (Lande and Arnold 1983 and Manley 1976, respectively). In this particular case, it is better to use logarithms because body measurements are usually log-normally distributed (for discussion see Sokal and Rohlf 1981). One must be certain that the assumptions of the methods are valid for the data. The normality assumption also affects discriminant functions (Lachenbruch and Goldstein 1979).

A major problem in multivariate analysis is missing variables. If an important major independent variable is not included in the study, then the apparent importance of the known variables will be inflated, and in some cases (depending upon correlations with the missing factor) it can even change the sign of the observed b'_k (Van de Geer 1971; Sokal and Rohlf 1981; Karson 1982). Body weight in the Bumpus data is a good example; its absence in the analysis makes humerus length apparently significant, which may be an artifact of phenotypic correlation with body weight. Every effort must be made to include all factors in any multivariate study, and the only way we can be sure of this is to know enough about the biology to know the *mechanism* of selection. This requires the use of methods IX and X in addition to VI–VIII (see Chapter 5). To make matters more complicated, as the number of variables is increased, it becomes more and more difficult to demonstrate significance of any one of them, even though the entire regression may be significant. Therefore, some optimum number of known functionally significant traits will have to be chosen for analysis.

Very large sample sizes are necessary to get reliable estimates of the b'_k or d_k, since only a few aberrant points can radically change the observed values, and with smaller sample sizes (< 100) the methods are extremely sensitive to adding or subtracting variables (Sokal and Rohlf 1981; Lachenbruch 1975; Lachenbruch and Goldstein 1979). The magnitude of this effect depends upon both sample sizes and b'_k magnitudes, but, since the b'_k are unknown, it is always safer to use as large a sample size as is possible. In addition, the total sample size must be larger than the total number of traits before it is possible to do the regression at all, and it must be considerably larger to yield significant b'_k (Sokal and Rohlf 1981; Lande and

Arnold 1983). The problem is particularly difficult in quadratic regression, or when variance-covariance matrices are heterogeneous: there the number of variables is a combination of all of the single and squared X_k values and cross-products.

Another important problem with multiple regression, especially quadratic regression, is that the estimation of the b'_k values is extremely sensitive to intercorrelation of the independent variables, which can lead to a singular $\mathbf{VV}'$ matrix, invalidating the estimates. This is a particular nuisance for few variables and smaller sample sizes. Of t strongly intercorrelated traits or variables, $t - 1$ should be deleted from the analysis, or, as O'Donald (1973), Lande and Arnold (1983), and others suggest, the analysis should be done using principal components or discriminant functions. Unfortunately, the biological meaning of principal components or discriminant functions is not always obvious.

6.4.6. *Illustrations of Multivariate Methods and Their Problems*

The best way to illustrate the methods and some of their problems is through examples of selection with known properties. Consider individuals that vary in two trait values, X_1 and X_2. Distributions of trait values were generated and various modes of selection were imposed on the simulated data (Appendix 2). Various measures of selection were tested on the results of selection. The *selection line* is the actual (imposed) line separating individuals with high and low fitness. The *separation line* is the estimate of the selection line from discriminant function analysis, and is algebraically related to the multiple regression of fitness on the two trait values (section 6.4.3). Details and results are given in Appendix 2, and summarized below.

Directional selection (Figure 6.2). The selection line is $X_2 = A + BX_1$, where A is the intercept and B is the slope. When B is at or near zero the contribution to fitness of trait X_1 is small; when B is very large, then X_2 contributes little to fitness. As expected, the magnitudes of b_k and b'_k parallel the changes in B; b_1 is small for small B, and b_2 is small for large B. Similarly, the separation line was usually a good estimate of the selection line, and this is also paralleled by the significance tests for regression coefficients (Appendix 2, Table A.1).

The accuracy of both the regression coefficients and separation lines declines with declining sample size. The accuracy also

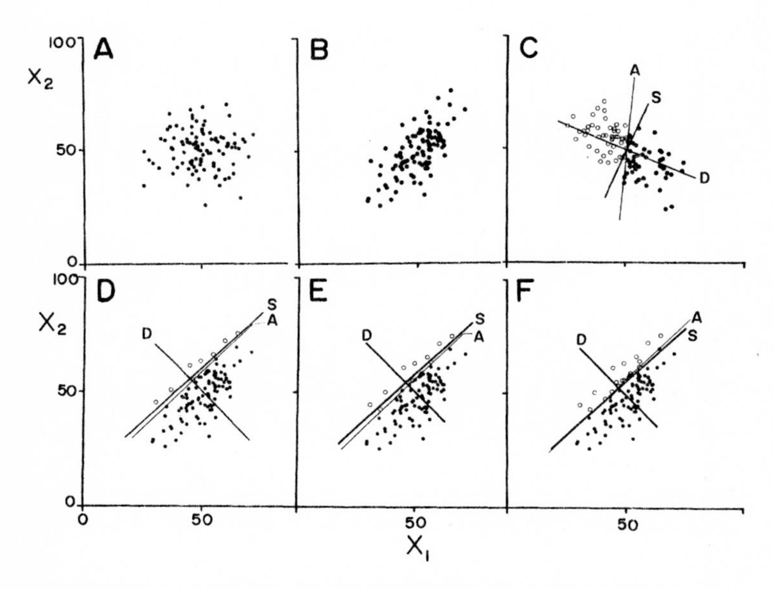

FIGURE 6.2. A plot of variation of 100 individuals with respect to two quantitative traits, X_1 and X_2, and four examples of errors in estimating directional selection functions. The phenotypic correlations between X_1 and X_2 are: *A*, -0.009; *B*, 0.693; and *C*, -0.664. In graphs *C*–*F* open circles may be regarded as either surviving or nonsurviving individuals. Line *A* is the actual selection line, the threshold separating live and dead. *S* is the estimated selection line (or separation line), derived from the discriminant function *D*. In graph *C* selection is nearly independent of an individual's X_2 value, yet the particular configuration of points yields an estimated separation line that contains a significant effect of X_2, as shown by the significant b_2 coefficient for X_2 in the multiple regression. Graphs *D*–*F* shows how accuracy of estimating the actual selection line increases as the line approches the grand mean; in other words, the estimate is more accurate for intermediate mean fitness $\bar{W}$.

declines with declining total population size and declining size of the smaller of the two groups. Sometimes, and clearly depending upon the chance position of individuals relative to the selection line, the separation and selection lines were significantly different. An example is shown in Figure 6.2C; although trait 2 had no effect on fitness, b_2 (and b'_2) is significantly different from zero, and the separation line has a much smaller slope than the selection line. The effect is more common when there is a significant phenotypic correlation between traits; stray points then have a greater effect, especially when the separation line is parallel to the principle axis of the phenotypic variation (Figures 6.2, 6.3). Over many generations of selection the cumulative effect becomes quite large, making both trait distributions drift much faster than that expected on the basis of effective population size and single trait population genetics alone.

The accuracy of the estimates are greatest for intermediate $\bar{W}$. As the selection line moves farther from the mean, the size of the favored or unfavored group declines, as does the sample size N_g, and chance positions of only a few individuals can greatly affect the separation line (Figure 6.2D,E,F; Table A.1). Strong asymmetries between N_a and N_d ($\bar{W} = N_a/N$) are similar to and affect the multiple regression in the same way as departures from normality. This inaccuracy affects both the slope and intercept of the separation line, though the effect on the intercept is greater. As a consequence, *there is a tendency for strong selection to be overestimated and weak selection to be underestimated.* Since the coefficients of the separation line are proportional to the multiple regression coefficients, these results, and those below, apply also to the multiple regression analysis of selection.

The effect of $\bar{W}$ on the accuracy of b_k and d_k is affected by phenotypic correlation. If there is phenotypic correlation among the traits, this effect is greatest when the slope of the selection line is parallel to the principal axis of the correlated traits (Table A.1). For example, if X_1 and X_2 have similar means and variances (as in the simulations), and if the phenotypic correlation is significantly positive, then the inaccuracies are maximized when the slope of the selection line is close to 1.0, and minimized when the selection line slope is close to −1.0. Equivalently, for multiple regression the inaccuracies are maximized when the values of the b_k are similar and of opposite sign, and minimized when the values are similar and of similar sign. If the phenotypic correlation is significantly negative, then the converse is true.

Directional selection with density dependence (Figure 6.3). Two examples are shown in Figure 6.3. Unlike the previous modes, there is now a significant nonlinear component to selection. The linear components behave roughly as they did for pure directional selection (Appendix 2, Table A.2), but there is greatly reduced accuracy in reconstructing the linear components. For

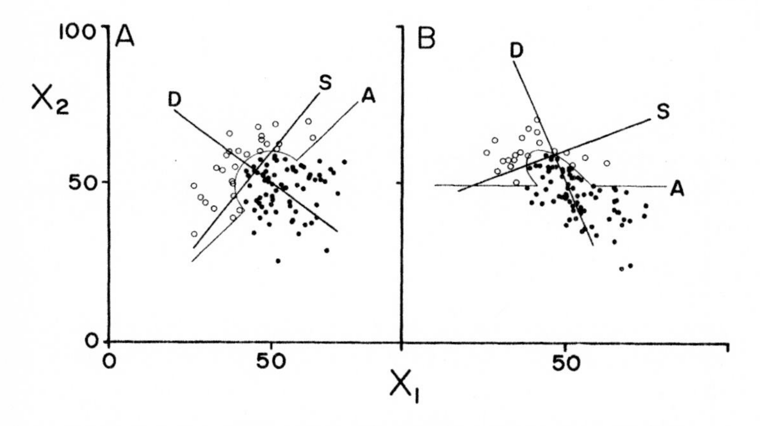

FIGURE 6.3. Directional and density-dependent selection of individuals that vary in two traits. Symbols as in Figure 6.2. Density-dependent selection (low density phenotypes at an advantage) was imposed by allowing individuals to die if their trait values were above a line defined by a given density contour on their bivariate normal distribution, and also above a directional selection line. Note that in graph *B* line *S* shows an effect of trait X_1 on fitness, and the multiple regression component b_1 is significantly different from zero; yet, except through density dependence, X_1 is not affected by the environment.

example, in Figure 6.3B there is actually no component of fitness due to trait 1, yet both linear *and* quadratic regressions yield a significant effect of trait 1 (Appendix 2, Tables A.2 and A.3). As for directional selection, this is due to the chance occurrence of particular combinations of traits, and is particularly common for strong positive or negative phenotypic correlations (compare Figure 6.3A and B).

Stabilizing or disruptive selection (Figure 6.4). As expected, the multiple quadratic regression in each case yielded significant quadratic terms. There is a serious problem in all multiple regressions including quadratic (and interaction or correlational selection component $X_k X_1$) terms in that some variables

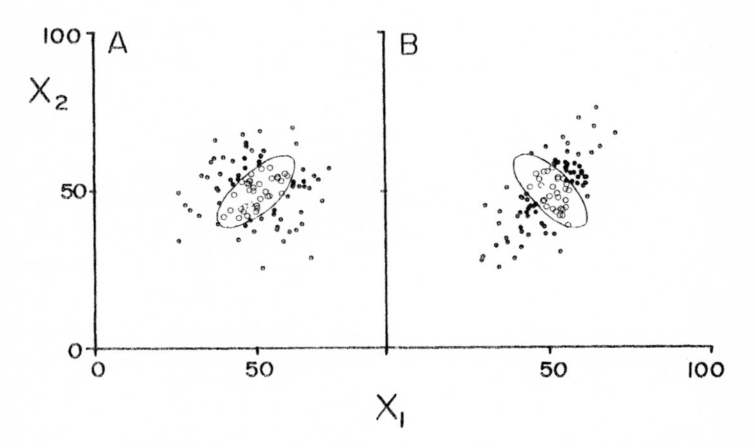

FIGURE 6.4. Variance and covariance selection, showing how correlations in the environment are not necessarily the same as the existing phenotypic correlations. In *A* there is no phenotypic correlation, but the environment favors a positive association between X_1 and X_2. In *B* there is a positive phenotypic correlation, but the environment favors a negative association. This significantly affects the variance-covariance matrices of the live and dead individuals. For example, the correlation between X_1 and X_2 for dead and live individuals is: *A*, -0.152 and 0.383; *B*, 0.764 and -0.399. The selection lines must then be estimated by quadratic discriminant functions, which do not assume equal variance-covariance matrices; in this case the selection "line" may be two hyperbolas or even an ellipse.

can be significantly and highly correlated. This results in the determinant of the $\mathbf{VV}'$ matrix being close to or actually zero (singular). If the determinant is very small, then unreasonable values of b'_k result, and their use in measuring and predicting selection becomes dubious. This can be avoided by stepwise regression. A strong correlation between linear and quadratic terms suggests nonnormality (Appendix 1). Figure 6.4A is also a good example of almost "pure" correlational selection, and 6.4B illustrates a mixture of correlational and stabilizing selection of both traits. See Arnold et al. (1986) for further discussion.

Assumptions. It is usually assumed that the variance-covariance matrices of the two groups are equal. This is at first sight reasonable, since the two groups were originally part of the same population. However, Figure 6.4 shows how the effect of disruptive or stabilizing selection of two traits does not have to yield the same correlation as the original phenotypic correlation between the traits. Since many generations of a particular form of selection would produce a particular phenotypic (and genetic) correlation, this would probably be uncommon in stable environments. However, in environments that change over tens or hundreds of generations, in environmental perturbations, or with invasions of new habitats, differences between correlated selection and phenotypic correlation may be found. In each case the variance-covariance matrices of the two groups (high and low W) are likely to be different. Thus an estimate of the selection line requires the use of the cumbersome quadratic discriminant function. The inequality of the $\mathbf{V}_g$ matrices is not limited to selection with nonlinear components. Depending upon where the selection line occurs, and especially if the line is far from the mean (large or small W) the groups can have quite different $\mathbf{V}_g$ from each other and from the original population. For example, a near-tangential slice of a pie far from its center (such as line S in Figure 6.3A) may result in a significant correlation in the slice, leaving the rest with $r = 0$.

The effect of various forms of selection of the variance-covariance matrices is shown in the correlation between X_1 and X_2 in Appendix 2 (Table A.4). For example, when $r = 0$, the group correlations can be significantly different from $r = 0$, and sometimes significantly different from each other. Changes in the environment are also likely to result in significant shifts in the form of the overall variance-covariance matrix, as has been experimentally demonstrated by Service and Rose (1985). One has to be very careful in analyzing and interpreting such data.

For both linear (directional) and quadratic (disruptive/stabilizing) components, if the assumptions are violated so that statistical tests cannot be made, the b'_k are still the best estimates of contributions to fitness by each variable (Lande and Arnold 1983). But then they are useful only to suggest avenues of further study.

Success of univariate measures. In spite of Lande and Arnold's (1983) warning that univariate measures of selection may give misleading pictures of the form of selection, they are surprisingly informative about the mode of selection: (1) the relative magnitudes of i and j (or i^* and j^*) indicate when there are directional and variance selection components; (2) i and j are highly correlated with the multiple regression coefficients; and (3) there is in general a good agreement between which traits are significant by univariate tests and which are significant on the basis of multiple regression (Appendix 2, Tables A.5 and A.6). As expected, the univariate measures do not perform as well as b'_k when there is significant phenotypic correlation among the traits. There tends to be an increase of larger i and significant j frequency if there is significant phenotypic correlation. Also, as expected, j' successfully corrects for directional selection on estimates of stabilizing/disruptive selection; it is less often large when there is only directional selection. The columns marked D in Table A.6 are calculations made on the trait values transformed onto the discriminant function. Note how they are in general about the same size as the larger of the two original variates if there is only one form of selection, but are much larger if there are two forms of selection, especially when the estimate of the form of selection is inaccurate.

It is clear that much more theoretical and empirical work must be done before we will understand how to measure natural selection on more than one trait at a time. This is extremely important because this is the way natural selection proceeds in nature.

6.5. THE USE OF SELECTION COEFFICIENTS AND DIFFERENTIALS

The inaccuracies in estimating the actual selection line do not affect the usefulness of the b_k' coefficients for predicting the response to phenotypic selection, since the response will be determined by who actually survives (or mates) to produce the next generation (Lande and Arnold 1983). In Figure 6.3B, the net effect of density-dependent and directional selection is to result in significant effects on fitness by both traits, even though the mechanism is dependent only on trait 1 through its density. Thus there is a significant difference between estimating selection coefficients (b_k') and using them to attempt to reconstruct why and how selection occurs (the form of the separation lines). Because the differences between the observed and actual selection lines can be so different, the b_k' coefficients are valid in predicting the response to selection in only one generation. A detailed knowledge of the *population* trait distributions and the average separation line is required for prediction for more than one generation.

The result of the difference between the actual selection line and the separation line (and associated b_k') can be a significant random fluctuation in the observed selection differentials among generations, as well as in the i^*, j^*, and b_k'. Even though the mechanism of selection may be constant (for example, a constant relationship as in Figure 6.3B), the b_k' may fluctuate a fair amount due to the chance combination of actual trait values in each generation. This may be significant even in populations that seem too large for ordinary genetic drift, and is expected even if we were omniscient and were able to sample the entire population each generation. Except for the resulting variance-covariance matrix, the consequences over many generations may be indistinguishable from genetic drift, as for random fluctuation of selection coefficients in polymorphic traits (Chapter 4). This needs more study.

There are at least five different uses of estimating selection: (1) to demonstrate its existence; (2) to estimate its rate; (3) to predict the genetic response in the next generation; (4) to understand the form of selection—why and how it occurs; and (5) to predict long-term changes in trait distributions. Only the simplest of tests need be used to demonstrate selection. Prediction of the genetic response to phenotypic selection requires much more extensive data from within a single generation, and can be obtained from the multiple regression coefficients. The standard errors of the multiple regression coefficients will give some idea of the accuracy of the predictions, but they will only be useful for that particular generation and its trait distributions. As a result of the vagaries of which individuals happen to be present each generation, the actual selection line may often be outside the 95% confidence limits from the regression, and, even if within the limits, may not be centered there. Thus it is necessary to obtain many generations of estimates of the selection line in order to get a good idea of the form and mode of selection. A detailed knowledge of the biological and ecological significance of the trait variance (methods IX and X) will greatly help in sorting out the form of selection. Only where a reliable and repeatable estimate of the selection line is made can it be used to predict many-generation change or (if possible) the predicted trait equilibrium. It is completely invalid to extrapolate to long-term changes or equilibrium from a single-generation estimate of selection on multiple traits, and this applies to single traits, too, because they are likely to be phenotypically or genetically correlated to other selected traits. This entails a tremendous amount of first-rate field work, but anything that should be attempted should be done properly.

As hinted in Chapter 5, it is a serious mistake to restrict a study to the estimation of selection coefficients and differentials. Carrying the chemical analogy (Chapter 2, section 2.2) further, this is equivalent to measuring the rate of a chemical reaction, and then not investigating the causes and mechanisms of the reaction. A highly accurate measure of selection differentials or coefficients, combined with a lack of knowledge of the reasons for and mechanisms of selection, is little more than refined alchemy. The causes and mechanisms must also be demonstrated with critical experiments.

6.6. SUMMARY

There are a variety of methods for estimating fitness and rates of natural selection, and these are summarized here. Methods for polymorphic traits primarily estimate fitness, while those for quantitative traits estimate the rate of selection within a generation—the rate of phenotypic selection. Methods for analyzing simultaneous natural selection on multiple traits are also summarized. They are able to eliminate the effects of phenotypic and genotypic correlations among traits and show how estimates of selection on single traits may be biased by correlations. Since the ability to survive, mate, and produce offspring is a function of all of an individual organism's traits, it is imperative that we consider more than one trait at a time. Work on methods for exploring natural selection on multiple traits is just beginning and should greatly increase our understanding of natural selection. A consideration of the problems of estimating selection on multiple traits shows that estimating selection for prediction of the genetic response is a significantly different problem from estimating how selection works, and is different from estimating the effects of many generations of selection. The difference also has some implications for apparent genetic drift. Random effects are greater for multiple-trait phenotypes than for single-trait phenotypes with the same N_e. To obtain reliable estimates of the multivariate selection coefficients b'_k, we must use data from many generations; single-generation estimates, even with large sample sizes, are very sensitive to chance combinations of traits in "outlier" individuals.

CHAPTER SEVEN

Distribution of Selection Coefficients and Differentials in Natural Populations

I know not any thing more pleasant, or more instructive, than to compare experience with expectation, or to register from time to time the difference between idea and reality. It is by this kind of observation that we grow daily less liable to be disappointed.
Samuel Johnson, 1758 (Boswell and Glover 1901, vol. 1, p. 220)

Many of the studies of natural selection listed in Table 5.1 present sufficient data to allow calculation of selection coefficients and differentials. From this we can obtain a crude idea of the distributions of selection coefficients and differentials in natural populations. Two earlier summaries are found in Antonovics (1971) and Johnson (1976), but these dealt with fewer than fifteen values, and so were unable to arrive at any conclusion about distributions. This chapter presents the distributions and compares them to roughly that which would be expected if selection coefficients and differentials varied at random.

7.1. METHODS

Selection coefficients and differentials were gathered from the literature cited in Table 5.1. Not all studies could be utilized because many authors published only reduced data or omitted some critical parameter (such as standard deviation or sample size). The species used are marked with asterisks in Table 5.1. In addition to these species, nine data points were included from a few single-locality-and-generation studies; these did not quite meet the criteria for admission to Table 5.1:

Parattetix texanus grasshoppers (Fisher 1939), *Uta stansburiana* lizards (Tinkle and Selander 1973), *Mus musculus* mice (Van Valen 1965b), and *Rattus rattus* rats (Van Valen and Weiss 1966). One nonsignificant datum from *Pogonomyrmex barbatus* ants was also included (Davidson 1982).

For comparative purposes, the same measures had to be used for all studies: the univariate measures S, i, and j' (see Chapter 6); multivariate data are presently too rare to be used; hopefully this will change in a few years. Because some of the univariate results may actually reflect selection on phenotypically correlated traits, they may underestimate the amount of direct selection; the more extreme values of the distributions may actually be conservative. The designs of statistical tests are also conservative (see below), and so the actual values of S, i, and j' may be larger than those observed; however, this measurement bias may be counteracted by the unknown frequency of studies that found no significant results and were not published. This will be discussed further in section 7.3.

Each calculated value was tested for significance; the null hypotheses are $S = 0$, $i = 0$, and $j = 0$, respectively. The selection measures and statistical methods were chosen so that as much published data could be used as possible, while still allowing the same methods to be used for all studies.

The selection coefficients S for polymorphic traits were tested as follows: Let a and b be the numbers of genotypes 1 and 2 before selection, and c and d be the numbers after selection. Then $W_1 = c/a$, $W_2 = d/b$. To make the maximum relative fitness equal 1, the other relative fitness w is either ad/bc or bc/ad, whichever is less than 1. The selection coefficient is $S = 1 - w$. Significance was tested by means of a 2 by 2 χ^2 test or Fisher's exact test when necessary (Sokal and Rohlf 1981). If there were more than two genotypes, then the largest W_k was used as the standard to calculate w_k for each genotype k, and the significance of each genotype's S was tested in a separate 2 by 2 test. If all tests were not significant, but the overall test was significant, then the largest S was regarded as significant and the others insignificant. This is conservative because it compares before and after rather than selected and unselected (see section 6.2.1).

The selection differentials i and j for quantitative traits were tested by means of equations 6.3 and 6.4 (Chapter 6). Smith's test (equation 6.5) could not be used for j because very few studies provided either raw data or data on kurtosis (which yield the required X^4 terms). To minimize the effect of directional selection on the variance, j' was calculated from i and j (equation 6.6). There are three ways to address the significance of j': (1) use the significance test for j; (2) use a given j' only when the associated i is not significantly different from zero; and, best (3), use the multiple regression tests (Appendix 1). Unfortunately, virtually none of the studies provided raw data on which it was possible to perform the multiple regression. However, it is possible to reformulate the expressions for the standard errors of the b_k (Appendix 1, equations A.10) in terms of i, j, the variance before selection, and the sample size. This allows an approximation to the multivariate tests (equation A.11). The standard errors are:

$$S_{b_1} = \sqrt{\frac{(N-1)\sum d^2}{N(N-3)v}} \tag{7.1a}$$

$$S_{b_2} = \sqrt{\frac{(N-1)\sum d^2}{N(N-3)2v^2}}, \tag{7.1b}$$

where

$$\sum d^2 = \frac{(1-\bar{W})}{\bar{W}} - i^2 - \frac{(i^2+j)^2}{2} \tag{7.2}$$

(symbolism as in Appendix 1). Using equations 7.1 and 7.2 and Appendix 1, equations A.11 and A.20, we could test i and j' in published data. Unfortunately, there are two serious problems. First, this test requires a reliable estimate of mean absolute fitness ($\bar{W}$ in equation 7.2); most publications give only means, variances, and sample sizes. Second, and most serious, this assumes normality. Even slight skewness tends to make the t values calculated (using equations 7.1 and 7.2) larger than those calculated on the raw data (using Appendix 1), and the bias is particularly bad in 7.1b. The result of the inflated t values is that i and j' will often seem to be significantly different from zero even when they are not, and the bias is as bad or worse than equations 6.3 and 6.4. For the purposes of comparison of published data, we will therefore have to be content with the simple tests of i and j (equations 6.3 and 6.4). I hope more people will use the multiple regression tests in the future, so that a more refined estimate of the distribution of i and j' can be made in a few years. The tests of the statistical significance of S, i, and j' are conservative since they test for differences between the before and after selection samples rather than between the selected and unselected samples; thus they will underestimate the frequency of significant values (see also Chapter 6).

The estimates come from a variety of life-history stages and intervals. Some estimate only a short period of viability selection, others a longer period, others estimate other components of fitness, and still others attempt to approach the total net fitness. Thus the fitness values themselves, even if they were perfectly precise estimates, are not directly comparable; none gives the total lifetime relative contribution to the next generation. For this reason the results given below should not be taken too seriously; they should be taken only to indicate in a general way the distribution of S, i, and j' .

The mean fitness measures (I and I_w) were also calculated. Negative values were quite frequent where j' indicated disruptive selection, indicating frequent violation of the assumptions of the selection models these estimates assume (Chapter 6). These measures were not related to significant or insignificant values of i and j in any obvious way; consequently they cannot be used for comparative purposes. This suggests that I measures should be used with caution. In addition, they are composite measures of both the effects of directional and variance selection, but not linear combinations of i and j'. For these reasons, the I and I_w distributions will not be reported here.

7.2. OBSERVED DISTRIBUTIONS

7.2.1. *Polymorphic Traits*

Figure 7.1 shows the distribution of observed S values, including those that were significant at the 0.05 level (shaded part of the histogram). The total frequency falls off with in-

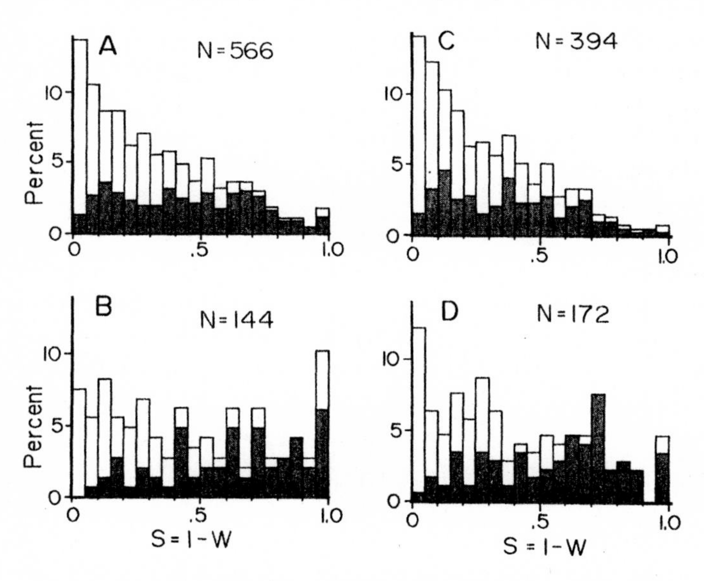

FIGURE 7.1. The distribution of selection coefficients S for polymorphic traits. The total height of each bar indicates the percentage of S values in each interval, and the shaded portion indicates the percentage of S that are significantly different from zero at the 0.05 level. *A*, data from undisturbed populations; 36 species, 239 of 566 S values significant. *B*, data from perturbations, field cages, or stressful environments; 12 species, 70 of 144 S significant. *C*, data from mortality selection; 34 species, 154 of 394 S significant. *D*, data from fecundity, fertility, and sexual selection; 13 species, 92 of 172 S significant.

creasing S, but there appears to be a uniform distribution of significant S.

There were enough data to subdivide the studies, and this revealed two interesting patterns related to disturbance and components of fitness. Some of the studies involved perturbations (method VI), including some in field cages (*Cepaea*). Other studies included stressful environments such as grasses on abandoned heavy-metal mines (for example, *Agrostis*). It is less likely that these populations are at equilibrium, compared to the relatively or actually undisturbed populations. Selection may be easier to detect and stronger if a population is away from equilibrium, especially if selection is frequency- or density-dependent. It is very clear that perturbed and stressed populations (Figure 7.1B) show larger S than undisturbed populations (Figure 7.1A). This may be evidence for widespread frequency-dependence. The data from Figure 7.1A were further subdivided into two components of selection: mortality (Figure 7.1C) and nonmortality (mating ability, fecundity, and fertility; Figure 7.1D). It is clear that nonmortality components of selection yield larger S than does survivorship, implying that they may be more important than survivorship in affecting polymorphic traits. It would be extremely interesting to know if this were commonly true within species. This lends some support to Lande's (1981) suggestion that sexual selection may be important to rapid divergence and speciation.

7.2.2. *Quantitative Traits, Directional Selection*

Figure 7.2 shows the distribution of i. The absolute values of i are shown since we are interested in magnitude rather than direction. As for S, the total frequency falls with increasing i, while the significant i are more uniform.

We can subdivide the i data (Figure 7.2) as we did for S. However, unlike S, the subdivision of the i data reveals no pattern. The Kolomogorov-Smirnov tests (Siegel 1956) between the groups were insignificant; there was no evidence for effects of disturbance and there were no differences among components of selection. The samples used to calculate the i were much smaller than those used to calculate the S (geometric means of 146 and 450, respectively), and there were many fewer i than S available to make Figure 7.2; but if the i results were as dramatic as in S, they might still have appeared. The suggestion that selection on a quantitative trait will yield very weak selection coefficients on the contributing loci (Milkman 1982; Lynch 1984) may also be relevant here. Much more work is needed to address this problem.

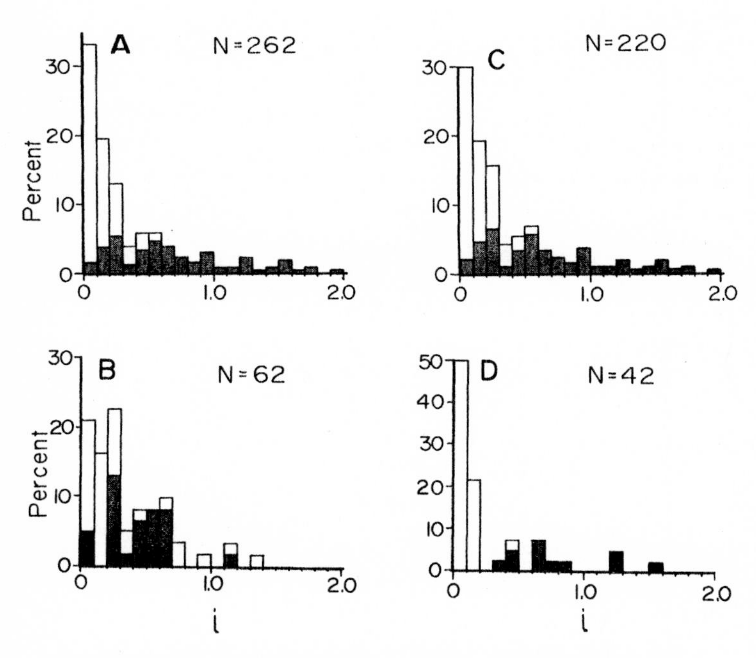

FIGURE 7.2. The distribution of directional selection differentials i. Only the absolute value of i is given because we are not interested in direction. Same conventions as in Figure 7.1. A, undisturbed; 25 species, 102 of 262 i significantly different from zero. B, disturbed; 5 species, 27 of 62 i significant. C, mortality; 17 species, 91 of 220 significant. D, nonmortality; 9 species, 11 of 42 i significant.

Data were also available for fossil and subfossil data from 5 species: *Agerostrea mesenterica* oysters (Sambol and Finks 1977), *Homo sapiens* (Perzigian 1975), *Merychippus primus* horses (Van Valen 1965a), *Ursus spelaeus* bears (Kurten 1957, 1958), and *Ursus arctos* (Kurten 1958). These data are shown in Figure 7.3. For all i, the fossil and subfossil i are significantly larger than i from living data (Kolomogorov-Smirnov two-tailed test, $P < 0.005$), but there is no difference between the distributions of *significant* i ($P > 0.05$). This pattern probably results because the data would not have been published if they had not shown a pattern, and especially because these data represent estimates from pooled effects of many generations.

It is interesting to note that the range of i values found in natural populations (Figure 7.2) extensively overlaps the values found in animal breeding and artificial selection experiments. For example, the i values found in Falconer (1981) and in the papers edited by Robertson (1980) range from 0.15 to 1.39, with a geometric mean of 0.71. For comparison, the geo-

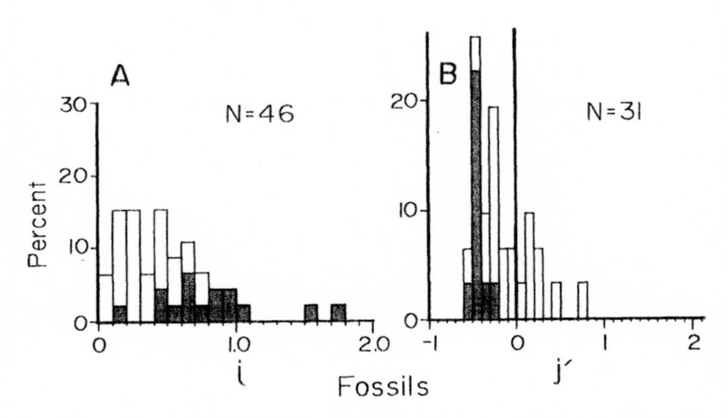

FIGURE 7.3. The distribution of directional and variance selection differentials in studies of fossil and subfossil data. Conventions as in Figure 7.1. *A*, directional selection; 5 species, 15 of 46 i significantly different from zero. *B*, stabilizing ($j' < 0$) and disruptive ($j' > 0$) selection; 5 species, 10 of 31 j' significant. *B* is a plot of all j' for which i is not significant.

metric mean of significant i in Figures 7.2 and 7.3 is 0.59. This suggests that natural selection is often as strong as artificial selection.

7.2.3. *Quantitative Traits, Variance Selection*

Figure 7.4 shows the distribution of j and j' for all data and Figure 7.3B shows the distribution of j' for fossil data. A negative value indicates stabilizing selection, and a positive value indicates disruptive selection. As for i, subdivision of the data by disturbance and selection components showed no pattern, and, in addition, there was no difference between the living and fossil distributions (Table 7.1).

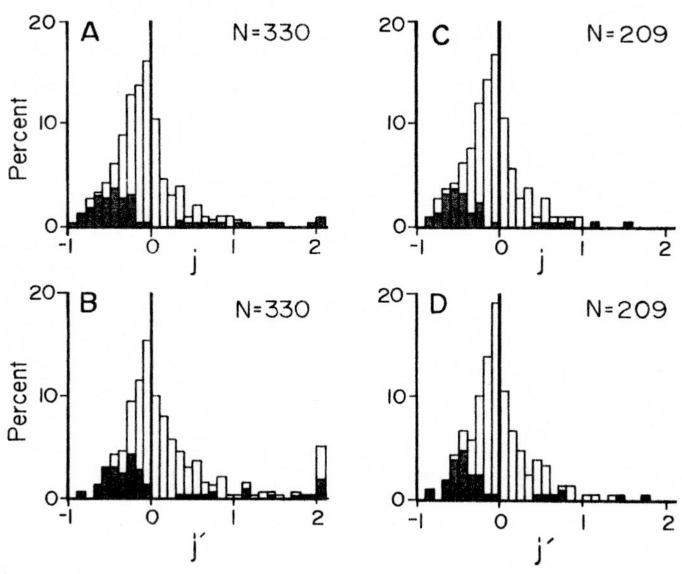

FIGURE 7.4. The distribution of variance selection differentials j and j'. A negative j indicates stabilizing selection (reduction in variance), while a positive value indicates disruptive selection. Shading conventions as in Figure 7.1. All data (including fossils of Figure 7.3); 32 species. *A*, distribution of j; 78 of 330 j significantly different from zero. *B*, distribution of j'; 78 of 330 j' significant. *C*, distribution of j when i not significant; 42 of 209 j significant. *D*, distribution of j' when i not significant; 42 of 209 j' significant. See text for discussion of significance tests.

Table 7.1 Distribution of all and significant j'

Class Limits of j'		All j'						5% Significant j'					
		Non-foss.[a]	Foss.	Un-dist.	Dist.	Mort.	Non-mort.	Non-foss.	Foss.	Un-dist.	Dist.	Mort.	Non-mort.
						Stabilizing Selection							
−1.0	−0.9	0	0	0	0	0	0	0	0	0	0	0	0
−0.9	−0.8	2	0	1	1	1	0	2	0	1	1	1	0
−0.8	−0.7	0	0	0	0	0	0	0	0	0	0	0	0
−0.7	−0.6	4	0	4	0	4	0	4	0	4	0	4	0
−0.6	−0.5	8	2	7	1	7	0	8	1	7	1	7	0
−0.5	−0.4	6	8	3	3	3	0	3	7	2	1	2	0
−0.4	−0.3	11	4	8	3	5	3	6	2	6	0	4	2
−0.3	−0.2	23	8	23	0	17	6	11	3	11	0	8	3
−0.2	−0.1	35	3	29	6	22	7	8	1	8	0	7	1
−0.1	0.0	48	3	43	5	34	9	3	1	3	0	3	0
Total		137	28	118	19	93	25	45	15	42	3	36	6
Differences[b]		$P < 0.005$		n.s.		n.s.		n.s.		n.s.		n.s.	

Disruptive Selection

0.0	0.1	32	1	24	8	17	7	0	0	0	0	0	0
0.1	0.2	21	5	20	1	17	3	0	0	0	0	0	0
0.2	0.3	15	4	9	6	7	2	0	0	0	0	0	0
0.3	0.4	14	1	9	5	9	0	1	0	1	0	1	0
0.4	0.5	9	1	6	3	6	0	1	0	1	0	1	0
0.5	0.6	11	1	9	2	8	1	1	0	1	0	1	0
0.6	0.7	4	1	4	0	4	0	1	0	1	0	1	0
0.7	0.8	3	1	3	0	3	0	2	0	2	0	2	0
0.8	0.9	7	0	7	0	7	0	0	0	0	0	0	0
0.9	1.0	1	0	1	0	1	0	0	0	0	0	0	0
1.0	1.1	1	0	0	1	0	0	0	0	0	0	0	0
1.1	1.2	5	0	5	0	4	1	3	0	3	0	2	1
1.2	1.3	0	1	0	0	0	0	0	0	0	0	0	0
1.3	1.4	2	0	2	0	2	0	0	0	0	0	0	0
1.4	1.5	2	0	2	0	2	0	1	0	1	0	1	0
1.5	1.6	1	0	1	0	1	0	0	0	0	0	0	0
1.6	1.7	0	0	0	0	0	0	0	0	0	0	0	0
1.7	1.8	2	0	2	0	2	0	1	0	1	0	1	0
1.8	1.9	1	0	1	0	0	1	1	0	1	0	0	1
1.9	2.0	1	0	1	0	1	0	0	0	0	0	0	0
2.0	+	15	2	14	1	12	1	6	0	5	1	5	0
Total		147	18	120	27	103	16	18	0	17	1	15	2
Differences		n.s.		n.s.		n.s.		n.v.		n.v.		n.v.	

[a] Headings (see also Figures 7.1 and 7.2): Nonfoss. = data excluding fossils. Foss. = fossil data. Undist. = undisturbed populations. Dist. = disturbed or stressful environments. Mort. = mortality selection. Nonmort. = nonmortality selection.

[b] Kolomogorov-Smirnov test, two-tailed (Siegel 1956); n.s. indicates $P > 0.05$ (not significant); n.v. indicates not valid to perform test.

Figure 7.4 shows the distribution of both j and j', each presented in two ways: the complete distribution (Figure 7.4A,B) and the distributions for cases where i is not significant (Figure 7.4C,D). Note that, although j' does correct for variance effects of directional selection (Chapter 6), it also tends to inflate the estimated frequency and magnitude of disruptive selection (Figure 7.4B,D); this suggests that j may be better than j' for measuring disruptive selection. Note also that eliminating j or j' when i is significant removes the extreme values (Figure 7.4C,D). The basic pattern is remarkably stable no matter how variance selection is measured. Insignificant small j and j' are most common and there is evidence for both disruptive and especially stabilizing selection of a wide range of magnitudes.

Significant disruptive selection was found in *Nucella lapillus* dog whelks, *Triturus vulgaris* newts, and *Uta stansburiana* lizards (in some cases in more than one trait); this was not discussed in the original papers except briefly by Bell (1974, 1978), who was more concerned with stabilizing selection. In addition, B. R. Grant (1985) found weak disruptive selection in *Geospiza conirostris* and Schluter et al. (1985) found it in *G. fortis* finches. Disruptive selection on quantitative traits favors phenotypic variation, though it will not maintain variation in the long term. As mentioned in Chapter 1, disruptive selection may arise from density-dependent selection because individuals closer to the mean will be more common, hence selected against. This might be happening in the first three species. Disruptive selection of bill shape and size in *G. conirostris* and *G. fortis* may be due to different processing efficiencies of distinctly different seed sizes (Schluter et al. 1985; B. R. Grant 1985). These are the first documented cases of disruptive selection in natural populations. Much more work needs to be done on disruptive selection and on methods of measurement. More field studies need to be conducted to determine the reasons for disruptive selection.

Bell (1974, 1978) considered and rejected developmental canalization as a possible reason for false stabilizing selection

($j < 0$) in *Triturus vulgaris*. The significant disruptive selection ($j > 0$) on some traits makes the canalization explanation even less likely, unless there are two alternative developmental pathways, which may yield false significant i and j of any value. The removal of j when i is significant affected some $j > 0$ as well as $j < 0$ for *Nucella*, *Triturus*, and *Uta* (compare Figure 7.4A,B with 7.4C,D). This is because there are strong directional selection components as well as a disruptive selection component in these species; the population mean may be closer to one "optimum" than another.

7.3. A COMPARISON OF OBSERVED AND EXPECTED DISTRIBUTIONS

Human nature and journal editors being what they are, it is difficult to publish data on attempts to measure selection that contain no significant values. Two exceptions are Jones and Parkin (1977) on *Cepaea* snails and Van Valen (1963b) on man. As a result, it is likely that the studies in Table 5.1 are a nonrandom sample, and the results in Figures 7.1–7.4 overestimate the frequency of large S, i, and j'. This may be partially mitigated by the use of conservative tests and methods that make acceptance of a significant value less likely (see section 7.1). The few studies that present data on all traits studied, including those with insignificant results, are therefore particularly valuable in indicating the distribution of selection coefficients and differentials (examples include *Triturus*, *Uta*, *Geospiza* and *Passer*). Although sample sizes are much smaller, the distributions are not different from Figures 7.2 and 7.4. Even these studies can suffer from correlations among traits. In order to estimate the actual distribution of S, i, and j', we need to know more about the frequency of insignificant measures. We can obtain a crude estimate by means of the expected distributions if S, i, and j' were random variables. These can be obtained from the chi-squared, t, and F distributions.

7.3.1. *Polymorphic Traits*

Consider two genotypes 1 and 2. Let w be the fitness of genotype 2 relative to 1. Let PT and $(1 - P)T$ be the numbers of genotypes 1 and 2 before selection, PT and $w(1 - P)T$ be their numbers after selection; assume no change in genotype 1. P is the proportion of genotype 1 before selection, T is the total before selection, $N - T$ is the total after selection, and N is the total sample size (before + after selection); $N = T(2 - S + SP)$, where $S = 1 - w$. Therefore the observed χ^2 value (with 1 degree of freedom) will be

$$X^2 = \frac{P(1 - P)S^2N}{2(2 - S)[1 - S(1 - P)]}. \tag{7.3}$$

From this and the χ^2 distribution, we can generate the expected distribution of S if S is a random variable. This yields a distribution similar in shape to the tops of the bars in Figure 7.1 (all S). It is difficult to test Figure 7.1 against this distribution because equation 7.3 depends strongly on N, and N varies greatly among studies. The mean and standard deviation of sample size (in natural logarithms since the sample-size distribution is skewed) are 6.11 and 1.15, respectively, yielding a geometric mean of $N = 450$. For large N the probability of S falls off rapidly with increasing S, while for small N it falls off more slowly. We can obtain a crude approximation of the composite distribution by drawing a large number of N from a random distribution with a mean and variance equal to the observed sample-size mean and variance, and substituting them into equation 7.3 and the χ^2 distribution. Comparing this composite distribution to Figure 7.1, we find fewer than expected small S values (significant + nonsignificant), especially below about 0.1. This discrepancy is probably due to the lack of publication of data with nonsignificant S.

Note that 309/710 or 43.5% of all the S in Figure 7.1 are significant, with a range of 39.1 to 53.5%. This is far more than the 5% expected by chance alone. The distribution of significant S is approximately that expected if all but the largest S values are equally frequent. This is because a larger sample size is required to detect a given smaller true S than a given larger true value. An estimate of the minimum sample size required to detect a given true S at the 0.05 level can be obtained by solving equation 7.3 for N using various P; this is shown in Figure 7.5. Note how it becomes progressively more difficult to detect a given S as S decreases. A different method of estimating N for a given S, which yields a similar relationship, is found in Hiorns and Harrison (1970). If the sample size is evenly distributed among studies estimating all values of S, we can expect the proportion of significant S to increase with S and have a shape similar to the shaded portion of Figure 7.1. A more refined prediction can be made with the observed mean and standard deviation of N of all studies. In this case, there are fewer than expected significant S values above about 0.75, either because the prediction is crude or because genotypes with large associated S have been eliminated from the populations. To be conservative, we can say that, given a large enough sample size, we do not expect any particular S magnitude in a natural

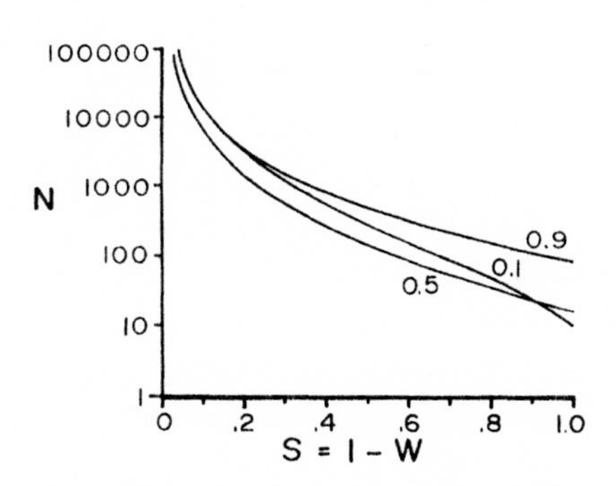

FIGURE 7.5. The minimum sample size (N) necessary to detect a true value of S for polymorphic characters; see text for discussion. The three curves are for different values of P, the proportion of the favored genotype before selection. Note the logarithmic N scale.

population. The condition for selection to overcome random genetic drift $S > 1/(4N_e)$ is easily met. Since virtually none of the data in Figure 7.1 comes from allozymes or other biochemical data (see marked species in Table 5.1), this conclusion may only apply to morphological and physiological traits.

This distribution is quite different from the distribution of fitnesses of mutations in laboratory populations of *Drosophila melanogaster* (summarized by Crow 1979). Considering all mutations, that distribution is bimodal, with many near-lethals, and many with small S. There are at least three reasons for the difference. First, as mentioned above, the S values estimate a variety of life-history stages. The data discussed by Crow were obtained and analyzed by a single method uniformly applied; they were therefore more likely to approximate the actual distribution of fitness values than the heterogeneous data in Figure 7.1. Second, a great diversity of traits and species were used to obtain Figure 7.1; the variation in their biology may be enough to mask any deficiency of intermediate S. Third, a major difference between laboratory populations of *Drosophila* and natural populations of any single species is that there is very little environmental variation among *Drosophila* bottles compared to variation among sampling sites of natural populations. It is possible that if natural populations lived in constant and uniform conditions, they too would have a bimodal distribution, but the variation among sample localities, and in time, would yield almost any value of S, depending upon the genotypes and the particular environmental parameters that happened to be present when the sample was taken. Since both genotypes and environment are varying among samples of natural populations of a single species, we expect (and observe) almost any value of natural S, whereas in laboratory populations of *Drosophila* with constant conditions, the distribution of S may be closer to the equilibrium conditions discussed by Crow (1979). Finally, lethals and near-lethals are less likely to be found in the field than in the laboratory. We must therefore use great caution in extrapolating from the lab to the field. The only general conclusion that can be drawn from Figure 7.1 is that selection coefficients greater than 0.1 are not rare, and are probably common in natural populations. This is in direct contradiction to the common assumption $S < 0.1$ of theoretical population genetics and many ecological geneticists (Ewens 1979; Wright 1978; Ford 1975), though it may still be safe for molecular or biochemical traits.

7.3.2. Quantitative Traits

An analysis similar to S can be made to obtain the distributions of i and j, which are related to the t and F distributions, respectively. Unfortunately, precise predictions of the distributions are still more difficult because more assumptions have to be made for them than for S. For i, assuming that N_a is approximately equal to N_b and $v_a = v_b$, $t = i\sqrt{(n/2)}$; the expected distribution of i can be obtained from this and the t distribution. For j, assuming that N_a is approximately equal to N_b, if $j < 0$ then $F = 1/(1 + j)$, and if $j > 0$ then $F = 1 + j$; the expected distribution of j can be obtained from these and the F distribution (j' can be obtained using both the t and F distributions).

The t distribution is strongly dependent upon the number of degrees of freedom ($2n - 2$ in this case), and therefore so is the expected distribution of i. The i distribution in Figure 7.2 deviates strongly from the i distribution derived from t with 144 degrees of freedom (from the geometric mean sample size). There are many fewer than expected small i, but also many more than expected large i. Again, this is probably a result of a tendency not to publish insignificant results (two notable exceptions are Davidson 1982 for *Pogonomyrmex barbatus* ants and Van Valen 1963b for man). Modification for unequal sample sizes within or between studies improves the fit slightly, but there is still the problem of assuming equal variances. The F distribution is strongly dependent upon the degrees of freedom ($N_b - 1$ and $N_a - 1$) and the normality assumption, so it is still more difficult to obtain a precise prediction of the expected distribution of j. As for i, and again assuming $N_b = N_a$, the shape of the observed j distribution (Figure 7.2) is similar to that expected, but again there is deficiency of insignificant small j.

45.4% of nonfossil i (Figure 7.2, range 26.2–43.5%) and 32.6% of fossil i (Figure 7.2A) were significantly different from $i = 0$, much greater than the 5% expected by chance alone. Similarly, 23.6% of j (20.1% of j') were significant, much larger than 5%. Figure 7.6 shows the minimum sample size necessary to detect a given true value of i or j at the 0.5 level, iterating $n = 2(t/i)^2$ for t as in Sokal and Rohlf (1981), and similarly for F. As for the polymorphic characters, a very large sample size is required to detect small i or j. Because of the dubious assumptions of roughly equal N and v, these graphs should be regarded as only a crude approximation; unequal N and v, which are expected in all selection studies (except some age-class comparisons), will *increase* the minimum sample size required to detect a given selection differential. In summary, given the observed mean and standard deviation of the sample sizes, we expect a distribution of significant i and j values similar in shape to Figures 7.1–7.4, but with more small insignificant i and j. Unlike S, there is no apparent excess of significant large i and j; this is not surprising because many loci contribute to the fitness of quantitative traits.

7.3.3. Conclusions

The general conclusion from the observed distribution of S, i, and j (Figures 7.1–7.4) is that one cannot say either that selection is weak or strong in natural populations, but rather that it can take any value, up to and including values found in artificial-selection experiments and in animal and plant breeding. The observed distribution of significant values is roughly uniform, with a deficiency in very small values because they require very large sample sizes for detection (Figures 7.5, 7.6). There is a suggestion that there is a deficiency of large S values because genotypes with large S (lethals and near-lethals) would tend to be rare in natural populations. It must be emphasized again that these are all partial estimates of fitness, in

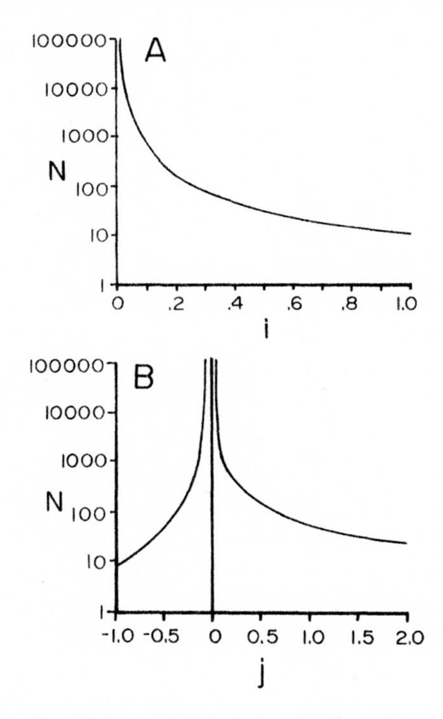

FIGURE 7.6. The minimum sample size necessary to detect a true value of i (A) and j (B); see text for discussion. Note the logarithmic N scale. These curves only suggest the relationship since they assume equal variances and approximately equal sample sizes (for i), or approximately equal sample sizes (for j). Unequal variances and sample sizes tend to increase the minimum N required to detect selection.

no case are they lifetime fitness estimates, and the form of the data and ecology of the species are quite heterogeneous. It is not reasonable to draw any conclusions from these distributions other than that strong selection is not rare and may even be common. Since strong selection may be common, at least for morphological traits (Table 5.1), it can easily overcome the effects of random genetic drift. The frequent statement that selection is usually weak in natural populations is without merit. For that statement to be true in the face of the large number of examples in Table 5.1 and Figures 7.1–7.4, there would have to be a tremendous number of unpublished insignificant studies.

I invite readers of this book to send to me their unpublished data on nonsignificant (as well as significant) selection measures. I hope to publish a revised distribution of field estimates in a few years, so any contributions or suggestions will be most welcome.

7.4. SUMMARY

Data were extracted from as many published studies in Table 5.1 as possible, and were used to show the distribution of observed selection coefficients and selection differentials. There is a large range of significant values from weak to remarkably strong. Polymorphic trait selection coefficients (S) above 0.1 are quite common, contrary to popular belief (and theoretical convenience). Directional selection differentials can be as high as in animal and plant breeding experiments. Stabilizing selection is less common than directional selection, and five species were found to show disruptive selection. From the distribution of nonsignificant selection coefficients and differentials, it is clear that there is a tendency not to publish nonsignificant results, so it is not possible, at present, to estimate the actual frequency distribution of selection coefficients and differentials in nature. A brief discussion is also made of the minimum sample size necessary to detect a given true S, i, or j. For polymorphic traits, though not at present for quantitative traits, perturbed or stressed populations appear to be affected by stronger selection than unperturbed populations, and mortality selection is less effective than mating ability, fertility, and fecundity. The frequent statement that selection is usually weak in natural populations is without merit.

CHAPTER EIGHT

The Importance of Natural Selection

> Learning confers so much superiority on those who possess it, that they might probably have escaped all censure had they been able to agree among themselves; but as envy and competition have divided the republick of letters into factions, they have neglected the common interest; each has called in foreign aid, and endeavoured to strengthen his own cause by the frown of power, the hiss of ignorance, and the clamour of popularity. They have all engaged in feuds, till by mutual hostilities they demolished those out-works which veneration had raised for their security, and exposed themselves to barbarians, by whom every region of science is equally laid waste.
>
> Samuel Johnson, *The Rambler*, 1 January 1751

It is clear from Chapters 5 and 7 that natural selection is ubiquitous enough to be found in a wide variety of organisms, and that strong selection is by no means uncommon in natural populations. Yet its importance to evolution, and to the maintenance of genetic variation, is a matter of opinion and extensive longstanding debate. Much of this debate is based upon varying emphases on different aspects of evolution by various people. The purpose of this chapter is to discuss these views briefly and to place them in perspective.

8.1. FOUR VIEWS

There are roughly four contrasting views on the importance of natural selection, and for brevity let us call them "selection," "random," "equilibrium," and "constraint." Some of these views partially overlap, and very few evolutionary biologists actually have views as extreme as any of them. Proponents of each view often downplay the other views in order to emphasize their own, but the names should only serve to focus attention on different evolutionary problems. A major difference between the views is their relative emphases on the processes of origin and change in frequency of new variants.

8.1.1. The "Selection" and "Random" Views

The "selection" and "random" views are common within the field of population genetics. They differ on the relative effects of selection and genetic drift on genetic variation as well as in evolution (people with these views are sometimes called, with execrable grammar, "selectionists" and "neutralists"). These views grew out of the older "balance" and "neoclassic" views discussed by Dobzhansky (1970), Lewontin (1974), Nei (1983), and Kimura (1983). This dialectic view and the associated selection versus genetic drift controversy have tended to obscure the multiplicity of views on natural selection and evolution.

The "selection" view proposes that natural selection is the most important factor in evolution (summarized in Mayr 1963; Dobzhansky 1970; Lewontin 1974; and Nei 1983). Genetic variability is maintained mainly by balancing, stabilizing, and geographically varying natural selection, which may shift as the environment changes. The effects of natural selection dominate other evolutionary factors. Genetic variation allows rapid response to changing environments. Mutation and recombination are the ultimate sources of this variation, but there is so much genetic variation that the rates and directions (if any) of evolution are set by selection "intensity." This view grew out of the older "balance" view, and unlike it, places less emphasis on mechanisms that maintain variation (such as heterozygous advantage) and more on directional selection and the *results* of the variation. For example, it is now known that balancing selection on quantitative or polygenic traits actually results in a loss of variation in the absence of mutation, yet selection is still regarded as very important in evolution (Lande 1976a, 1980).

The "random" view proposes that natural selection has a minor role and that randomness, caused by mutation, recombination, population structure, and genetic drift are much more important, even playing the "creative" role in evolution (reviewed by Lewontin 1974; Nei 1983; Kimura 1983). A substantial proportion of mutants or variants are selectively neutral or nearly neutral. Many mutants are deleterious, and only a small proportion are advantageous, but the latter are sufficient for evolution by natural selection. Sometimes morphological traits may be subject to changes in evolution that are not due to natural selection. Natural selection "purifies" by removing deleterious mutants and saving beneficial ones. New mutations spread through populations by natural selection or genetic drift, but a large proportion of them are eliminated by chance. Populations do not necessarily have the genetic variability needed for rapid response to natural selection and changing environments. If there is insufficient variability, the population remains unchanged until new mutations arise or until it goes extinct. Mutation limits the rates and directions (if any) of evolution. This view grew out of the older "classical" view, but places more emphasis on the dynamics of selectively neutral traits and mutation, and less on "purifying" selection. This view has also shifted its attention largely to variation and evolution at the molecular level, and allows selection to be important on morphological traits (Kimura 1983). For an attempt to reconcile neutrality of molecular traits with selection of morphological traits, see Milkman (1982) and Lynch (1984).

To clarify the "random" view's suggestion that mutation is the major factor in evolution, we can carry Ghiselin's (1981) erosion analogy further (Chapter 2, section 2.2). If natural selection is analogous to erosion, then mutation is analogous to orogeny (mountain building), and to a great extent the pattern of erosion will be determined by the particular kinds of rocks that make up the mountains. Erosion is slight on a plain, as is natural selection when genetic variation is minimal, but it is active in areas of high relief. It is myopic to say that the most important factor in evolution is *either* natural selection *or* mutation and genetic drift, just as it is wrong to say that topography can form with only orogeny or only erosion—both are absolutely necessary.

On the basis of present evidence, it looks as though the "random" view is relatively more successful in explaining allozyme and other biochemical and molecular variation. The bulk of the data is consistent with the predictions of "neoclassic" models (method v in Chapter 3; Nei 1983; Kimura 1983; Mukai and Nagano 1983), and there are relatively fewer demonstrations of natural selection on allozymes (Chapter 5). However, one has to be very careful because consistency is not proof (ironically, as Gould and Lewontin 1979 have argued about selection), and there is so little known about the function of allozymes. In addition, the match between the prediction of the relationship between the heterozygosity or the effective number of alleles and the product of effective population size (N_e) and mutation rate (u) is very poor; there is less heterozygosity than expected for large N_e (Nei 1983). Nei showed that although the match is poor, when the test is done properly the fit to a "selection" hypothesis is even worse than the "random" hypothesis fit. The mismatch results from overestimates of N_e, but correction of the sources of bias in estimated N_e is not enough to correct the graphs to yield more heterozygosity than expected, which would happen if the "selection" model were true. Milkman (1982) and Lynch (1984) discuss additional reasons for apparent neutrality but conclude that this results because molecular traits are contributors to quantitative traits; effective selection on quantitative trait polygenes can be very small even if i is large on the trait as a whole. This needs more work.

A frequent criticism of the "random" view, and one explanation for the abundant allozyme variation, is that the geographic ranges of most species include great environmental heterogeneity, so there is probably enough geographically varying (but not necessarily balancing) selection to maintain the polymorphisms in the face of gene flow. A variety of models (Ewens 1979; Endler 1977), and some extensive data from nattural populations of *Drosophila melanogaster* (Mukai and Nagano 1983) support this idea. A frequent counterargument is the fact that deep-sea organisms that have been examined by electrophoresis show as much variation as do other organisms. This assumes that the deep sea is more temporally and spatially constant than other habitats (see Valentine 1976). However, recent oceanographic work has shown this assumption to be quite false on both large and small scales (Gardner and Sullivan 1981; Wimbush et al. 1982). Since there is a good fit with most predictions of the "random" view, though there have been very few serious efforts to learn the function of molecular variants (Chapter 5), and because the heterogeneous environment model is a reasonable one, the problem of the maintenance of extensive protein variation is still completely unresolved. There is no such difficulty for morphological and many physiological traits, where there is abundant evidence for natural selection (Chapters 5 and 7). Although there are few extensive studies of lifetime fitness, the observed broad distribution of S, i, and j (Chapter 7) suggests that the effects of random genetic drift can be easily and frequently overcome.

Lewontin's (1974) discussion of the "balance" and "neoclassical" views or "schools" stresses the maintenance of genetic variation by natural selection or by mutation and genetic drift, respectively. The discussion of the views in this book emphasizes not the maintenance of variation, but the various factors that influence the rates and directions of evolution (changes in variation). The importance of natural selection in the maintenance of variation and in evolution in general are two very different questions. This is because (1) natural selection does not necessarily result in evolution (Chapter 1), and (2) the basis for long-term evolution is not only a change in allele frequencies (or trait distributions) or maintenance of variation, but also origin and replacement of some variants by others. I will use the term "replacement" to indicate any change in frequency of a trait value or allele from near zero to 100%. It is important to make a distinction between replacement and the substitution of an allele or base at a locus or in a DNA sequence during a mutation event. Confusingly, the word "substitution" has been used in the literature for both processes, but one involves an individual and the other a population or species.

Natural selection affects the frequencies of alleles, and if there were an infinite amount of genetic variation present, it would determine the rate of replacement of alleles, along with genetic drift; the theoretical rate for a quantitative trait is $N_e i$ (summarized in Hill 1982a,b). But most selection experiments, including animal and plant breeding, have found plateaus or limits to the response to selection in experiments lasting more than thirty to fifty generations (Falconer 1981; Robertson 1980). Only four cases are known in which no plateaus were observed over a large number of generations (Hill 1982a). After many more generations, a few of the experiments that exhibit plateaus start to respond to selection again, followed by a new higher plateau (an example is Frankham's paper in Robertson 1980). Plateaus are only eliminated over time, presumably by recombination or mutation. It would be very interesting to know what proportion of the release from the plateaus was due to mutation and to recombination; but in any case there is a limit to how fast a population can respond to long-term selection. This suggests that natural selection alone may not explain the tempo of evolution, and that mutation rate is very important.

However, there are several objections to plateaus as indicators of the limits to variation. Hill (1982a) gives four explanations for plateaus that do not involve a loss of genetic variation: (1) Responses to selection may be so small that they cannot be detected statistically. (2) Only deleterious mutants have occurred, and they are then lost. (3) A physiological or biophysical limit or other constraint may have been reached; all possible positive mutants have been fixed and any new mutants are deleterious. (4) An equilibrium or quasi-equilibrium may be reached between different components of natural selection (in an artificial selection experiment, natural selection may oppose artificial selection; an example is Enfield's paper in Robertson 1980). In other words, a plateau is either a trivial or nontrivial equilibrium. In addition, natural selection is specific to particular environments (by definition), and environments change in space and time, so plateaus may be artifacts of relatively constant environments. This and the ubiquity and magnitude of natural selection (Chapters 5 and 7) suggest that natural selection can explain much of the mode of evolution. Mutation and random genetic drift may be sufficient to explain evolutionary rates, but selection may be more important in considering the mode or direction of evolution. Little work has been done on natural populations that would address this problem, and such work is badly needed.

The existence of plateaus and the results of Chapter 7 bring up an inconsistency in both the "random" and "selection" viewpoints. According to the "random" view, selection is generally weak; therefore it will take a long time for selection to reduce additive genetic variation and produce plateaus. In that case the rate of evolution may be set by selection and *not* by the mutation rate. Thus the "random" premise does not necessarily predict that evolutionary rates are proportional to the mutation rate. An additional problem is that we can no longer assume that selection is usually weak (Chapter 7). According to the "selection" view, selection is generally strong; therefore additive genetic variation will be lost relatively quickly, except for that maintained by mutation, selection, pleiotropic, and epistatic effects (Ewens 1979; Hill 1982a,b; Lande 1976b, 1980). This may limit the rate of evolution to the mutation rate. Thus both extreme viewpoints are logically inconsistent and both mutation *and* selection are important in evolution.

There is yet another inconsistency in the literature. The proponents of the "punctuated equilibrium" model of evolution (Stanley 1979, 1982; Eldredge and Cracraft 1980) and the critics of the so-called "adaptationist program" (Gould and Lewontin 1979; Gould 1980, 1982) criticize the "selection" view in a way which suggests that they confuse it with the "random" view: they say that the "selection" view assumes *weak* selection, and this gives rise to constant evolutionary rates. Perhaps some people with more extreme versions of the "random" view would agree; but most would not predict constant rates of evolution—only average rates that depend upon mutation rates and N_e, and N_e is expected to change in time. In any case, the finding that there is a broad spectrum of selection (Chapter 7) implies that there is a broad range of relative effects of selection, drift, and mutation. This would suggest that there is also a broad range of rates of evolution. It is clearly unwise to state categorically that rates of evolution are bimodal ("punctuation") or unimodal with a small variance (the "punctuationists'" parody of "gradualism"). A frequency distribution of rates from the fossil record is not yet published, and the results of Chapter 7 imply almost a uniform distribution. It is therefore unreasonable to make broad generalizations about the relative importance of selection, genetic drift, and mutation in evolution.

8.1.2. The "Equilibrium" View

The "equilibrium" view is moderately common outside population genetics, and reaches its best development in behavioral ecology. It is similar to the "selection" view in that it assumes that selection is the major factor in evolution. Like the "selection" view, the "equilibrium" view assumes that there is enough genetic variation to overcome the effects of genetic drift and to allow rapid evolution. However, there are three important differences: (1) The "equilibrium" view assumes that natural populations are at or very close to equilibrium, and makes predictions on the basis of population models at equilibrium. The implicit assumption is that the environment has remained constant long enough for the equilibrium to have been reached. (2) It makes no explicit assumptions about the genetics of the traits of interest, and assumes that there is enough genetic variation to attain any phenotype, within the constraints of the models—hence, to attain any predicted equilibrium. (3) It makes explicit assumptions about various selective factors to derive models that are based upon general biology rather than purely arbitrary fitness models. The basic question of the "equilibrium" view is: "Can we understand enough about the internal trade-offs (size relations, physiology, biomechanics, developmental pathways, etc.) and the external trade-offs (predators, competitors, pathogens, other members of the same species or sex, seasonality, etc.) to calculate where the equilibrium will be?" (Charnov, pers. comm. 1984). Another formulation is: Are there any general rules that can explain observed variation and evolutionary patterns? Excellent examples and discussion are found in Charnov (1982) and Bull (1983). This view is the basis for method x for demonstrating natural selection (Chapter 3, section 3.11).

There are two major causes for concern: the assumptions about genetic variation and the assumption that populations are at or near equilibrium. If the predicted value is not constant, then the population or species can never have enough time to reach equilibrium, even if there were enough genetic variation to do so. We do not know enough about genetics or ecology to know how often any of the assumptions are met. To make matters more difficult, most models arising out of the "equilibrium" view yield frequency-dependent fitnesses, which tend to be hardest to detect when the population is at or near equilibrium (Chapter 6; Ewens 1979; Hedrick 1983). On the other hand, frequency-dependent selection appears to be common (Chapter 5), perturbed populations seem to yield higher selection coefficients than unperturbed populations (Chapter 7), and this viewpoint has had some remarkable success in predicting trait distributions (Charnov 1982).

An interesting sidelight to the "equilibrium" view is that the equilibrium may not be represented by a single point. Multiple equilibria are well known in frequency-dependent systems (Clarke 1975; Ewens 1979). More important still is that there may be a very large number of equilibria in any one system. For example, "lines" of equilibria are common in reproduction-related systems: sexual selection (Lande 1981; Kirkpatrick 1982), sex determination systems (Bull 1983), and plant-pollinator systems (Kiester et al. 1984). Lines of equilibria are series of combinations of two (or more) trait values toward which the population tends to evolve. A population may move along the equilibrium lines by genetic drift or by selection of correlated traits. In addition, in sexual selection the population may exhibit "runaway" behavior (Fisher 1930; Lande 1981; Kirkpatrick 1982), causing rapid shifts of trait distributions that would not be predicted by simplistic equilibrium analysis. The "lines of equilibria" phenomenon has only recently been discovered, and its full implications have probably not yet been fully appreciated. They are easily incorporated into the "equilibrium" view (Bull 1983), and so offer no major problems to the associated theory. However, environments usually change, the importance of nonmortality components of selection may have been greatly underestimated (Chapter 7), and lines of equilibrium may arise from these components; therefore the assumption that an equilibrium is necessarily a static point should be used with caution.

This "equilibrium" view grades into the "selection" view, and there are many intermediates. They range from those that merely relax the equilibrium assumption, to those that make explicit assumptions about genetics. The general approach associated with these intermediate views is discussed in Chapter 3 under method IX (section 3.10). The "equilibrium" view also grades into the "constraint" view.

8.1.3. The "Constraint" View

Many researchers distrust the assumptions of the three preceding views and are worried by the frequent lack of attention paid to the complexity of genetic systems and to the mechanisms of development. This has led to the "constraint" view. Two major ideas constitute the "constraint" view: (1) All mutants are not equally likely and depend upon the developmental and genetic systems, which themselves depend upon phylogenetic history. (2) Genes and traits do not work as independent units but interact in various ways and at various levels. These factors place constraints on what sorts of allelic substitutions are possible during evolution, and hence affect the rate and direction of evolution. This view dates back to Darwin (1859), and has been considered by a few evolutionary biologists[1] for some time, though it has been largely neglected until recently.[2] Raff and Kaufman (1983) provide a most interesting and useful discussion of this view.

The "constraint" view can be clarified by another extension of the erosion analogy. Mountains are not homogeneous masses

[1] See Cain 1964; Fisher 1930; Wright 1931, 1942; Mayr 1963, 1983; Williams 1966.

[2] See Gould 1980, 1982; Gould and Lewontin 1979; Eldredge and Cracraft 1980; Stanley 1979, 1982; Raff and Kaufman 1983, and references therein.

of rock; they are extremely heterogeneous: different parts have bedding planes oriented at different angles, and they differ in both mean and variance of hardness. This affects and constrains the effects of erosion as well as orogeny. For example, the orientation of the bedding planes of the Appalachian mountains run primarily northeast and southwest, and the dip is very steep in any one place. Many of the streams draining the complex follow the weaker rock strata and thus run parallel to the bedding planes—hence the main mountain axis—before leaving the system. This is a result of the geological history of the mountains and would not be expected if the mountains were homogeneous, which would yield randomly oriented streams. Thus not only the history of the material making up the mountains, but also orogeny and erosion determine the shape of the Appalachians. It is insufficient to consider only one factor.

Constraints include limits to variation set by phylogeny, physiology, biophysics, genetics (pleiotropy, epistasis, linkage, etc.), development, functional relationships, and the ability of development and physiological processes to assess the environment; these are obviously interrelated. All can constrain natural selection and evolution, give rise to plateaus, or bias future evolutionary changes in certain directions (reviewed by Charlesworth et al. 1982; Slatkin 1983; and Raff and Kaufman 1983). Natural selection cannot proceed when genetic variation is exhausted, or if there is insufficient time for useful mutations to accumulate. Extensive gene flow can also counteract selective effects (Endler 1977). In addition, when one gene affects two or more traits, or when significant genotypic correlation occurs among traits, and there is selection on one of them, the response to selection will be limited by the extent of selection on the correlated traits.[3]

[3] See Bulmer 1980; Falconer 1981; Slatkin 1983; Dobzhansky 1956, 1970; Wright 1931, 1942, 1965, 1982.

Constraints have fascinating implications for subsequent evolution. Consider the lack of cellulase in most herbivores and the lack of photosynthesis in animals. Both constrain subsequent evolution in certain directions. The presence of either of these systems in many animals would be very advantageous, yet it has happened rarely (cellulase) or not at all (photosynthesis). A few groups of animals have "solved the problem" by means of symbiotic organisms that do possess the proper devices—for example, zooxanthellae of corals and flagellates in termites. The accident of possessing or not possessing a particular biochemical system or mechanical device can have major effects on the subsequent evolution of the groups possessing the useful contrivance. Possession of the contrivance allows the utilization of new foods or other resources and invasion of new habitats, and it may shape the mode of subsequent morphological and physiological evolution.

Another example is the early evolution of internal (vertebrate) or external (arthropod) skeletons. Currey (in Cain 1964) shows that below a certain size the skeletal muscles may be so small relative to the skeleton that they can be accommodated within hollow tubes; but as size increases, physical scaling and energetic relationships favor an internal skeleton. For animals that have to move quickly or perform complex movements, there is an advantage to having an exoskeleton in small forms or an endoskeleton in large forms. The initial evolution and adaptation to those two modes of life have major implications, constraints, and biases on subsequent evolution with respect to size, respiration, water balance, and so on. For example, the slow-contracting parietal muscular systems of *Metridium* anemones consist of a single fiber thickness. It has been shown that they are as efficient as would be a rapidly contracting fourteen-layer muscle of the same shape (Cain 1964). The initial evolution of anemones in this direction may have prevented them from later evolving fast-moving forms with hard skeletons, and a similar argument may be made for the early dichotomy between hydrostatic and jointed skeletons (Cain 1964). These examples involve multiple adaptive solutions that were evolved a long time ago; but it is quite likely that similar, if smaller-scale, differences and constraints are evolving continuously.

In the other views, alleles were usually assumed to be sufficiently independent (even *with* epistasis, pleiotropy, or intercorrelations)—that there was no theoretical reason for irreversibility or directionality in evolution, given mutation and enough time. This is not assumed in the "constraint" view and is also rapidly becoming a less common assumption of the "equilibrium" view. For example, there are now good theoretical reasons for directionality caused by particular combinations of traits from which it is difficult to diverge; examples are Muller's ratchet, polyploidy, parthenogenesis, dioecy, selfing, haplodiploidy, and sex chromosomes (Bull and Charnov 1985). In addition, the trade-offs of the "equilibrium" view are explicit assumptions about constraints to variation. The "equilibrium" view assumes that the constraints actively determine or affect the equilibrium, while the "constraint" view merely regards the constraints as passive limits to variation whether or not a species is at equilibrium. The difference is analogous to the difference between population regulation and control, respectively. Thus the "equilibrium" and "constraint" views are converging with regard to the effects of constraints and directionality of evolution.

Note that by directionality, the "constraint" and "equilibrium" views do not suggest orthogenesis by a supernatural force, but merely that the possession or absence of a particular structure can bias or constrain subsequent evolution in certain directions. A related idea is that, as a result of interacting biochemical or developmental networks, some mutations are more or less likely to be expressed. Depending upon the interactions and the location of the mutation(s) in the network, rather sudden morphological and evolutionary change may happen (Gould 1982; Raff and Kaufman 1983). The key difference between the "constraint" and other views is a major concern about the *origin* of new variants rather than the mechanism of population equilibrium or replacement of some variants by others. Another important difference is that the "constraint" view emphasizes major morphological changes rather than predictable trait configurations. These are interesting and almost entirely unexplored ideas, and well worth thinking about (see especially Raff and Kaufman 1983).

The correlations and interactions among genes may themselves be affected by natural selection, through the selection of modifiers (Fisher 1930; Wallace 1981; Endler 1977; Slatkin 1983), or directly on the genetic systems concerned. Linkage among loci can have similar results (Slatkin 1983). These factors have had little investigation and will repay further study. A good example is rodent allometry. Atchley (1984), Atchley and Rutledge (1980), and Leamy and Atchley (1984) have shown that the allometric relationship among skeletal traits is itself heritable and can change over many generations of selection.

Less well explored are functional relationships. If two or more traits are functionally or developmentally related, or one trait serves several independent functions, the selection on one will affect or constrain changes in the others (Slatkin 1983; Falconer 1981; also Stanley 1979, 1982; Gould 1980, 1982). Wright's (1931, 1942) shifting-balance theory suggests that it may be difficult for a species to move between well-adapted genetic systems or into new adaptive zones that require a low-fitness intermediate (see also Mayr 1963 and Simpson 1944, 1949). A simple example is a tasty insect which mimics one species of a distasteful model insect, but which is genetically capable of mimicking a second, even more distasteful species. In order to evolve the second color pattern it would have to go through maladaptive intermediate stages that confer no mimetic protection. If the conditions for the shift are not met (small population size, etc., see Wright 1942, 1982), the species will continue to coevolve with the first model, and its subsequent history would be very different from that expected if simple ("global") optimization were possible.

Although there is little critical direct evidence for or against constraints at this time, it is safe to say that they will be important. But it is unreasonable to say that they are more important than natural selection, mutation, or genetic drift; one or all must be present to cause evolutionary change—with or without constraints. As Cain's (1964) and Atchley's (1984) examples show, the constraints themselves must have evolved and may themselves have resulted from natural selection. If the ancestors were polymorphic for developmental pathways, the heritably more efficient pathways would probably have spread at the expense of the less efficient. This has not yet been investigated and would repay further study.

8.1.4 Other Views

There are also some nonscientific reasons for regarding natural selection as unimportant. The distribution of observed instances of selection (Chapter 5) helps to explain a sociological observation: those who think that natural selection is the most important factor in evolution work primarily with morphological (and some physiological) traits in natural populations, and those who consider natural selection to be unimportant tend to work with molecular or biochemical traits in laboratory populations. If the distribution of observed instances of selection is accurate, then both groups are right, when considering their respective traits and working conditions. But both are also wrong in generalizing to other traits, because the prevalence of selection appears to be unequal among morphological, physiological, and molecular traits.

Others believe that natural selection is unimportant for political reasons. Almost without exception, they do not work with natural populations, though they may make brief sorties for collecting. Some phylogenetic systematists fall into this category ("cladists"; see especially Platnick 1977; Rosen 1978; Rosen and Buth 1980). The problem they face is that natural selection causes convergence ("homeoplasy"), which obscures the phylogenetic information that random divergence would otherwise provide. By minimizing the importance of selection, some cladists make a consideration of the choice of systematic traits, and the general validity of their methods seem less of a problem (discussion in Endler 1982a). The majority of cladists are much more thoughtful (for example, Wiley 1981), and are well aware that much more work must be done on the effects of natural selection on phylogeny reconstruction. The problem of the effects of natural selection on phylogeny reconstruction is an interesting one and would repay further study. Another group, consisting mainly of paleontologists and others holding the "constraint" viewpoint, have also denigrated natural selection to draw attention to their own views; a sort of intellectual lek (Stanley 1979, 1982; Gould 1980, 1982; Gould and Lewontin 1979; Eldredge and Cracraft 1980). They have often been too extreme in an effort to make people take notice, but they have stirred up serious interest in evolutionary constraints after a long period of neglect.

8.2. ORIGIN AND REPLACEMENT

There are many good biologists who think that natural selection (and all of population genetics) is almost, or in fact is, irrelevant to evolution. Because many are molecular biologists, their view might be called the "molecular" view, but there is a broad overlap with both the "random" and "constraint" views. At first this viewpoint seems puzzling. But it is understandable when one realizes that they see evolution mostly in terms of origin of new alleles or sequences, and deemphasize the process of population frequency change, or replacement. To them it does not particularly matter what happens during the replacement, or in some cases even why the replacement occurs. They are primarily interested in the *origin* of the variants that make the replacements possible. The mechanisms of mutation are in their viewpoint the most important factor in evolution, since at the most basic level, evolution is not possible without mutation (or recombination). This is an entirely different question from the causes of replacement once the variants have appeared. It is the most basic of questions about evolution because a fundamental understanding of the origin of new variants would allow us to address how morphological and genetic changes actually take place, as well as how they affect the rate and direction of evolution. Natural selection only affects changes in the frequency of the variants once they appear; it cannot directly address the reasons for the existence of the variants. This appears to be the fundamental position of those holding the "constraint" viewpoint.

The differences among the four principal views can be summarized by means of a diagram (Figure 8.1). Consider a hypothetical species that varies in phenotypic value with respect to two traits, 1 and 2. Any individual can be represented as a point in the (X_1, X_2) space. (A real organism could be represented as a point in n-dimensional space defined by n traits.) A population can be represented by a cloud of points; in Figure 8.1 the cloud is simplified to an ellipse (marked P), where the degree of ellipticity is proportional to the degree of phenotypic correlation between the traits. The range of values found in a population is a subset of the actual range found in a species (marked S), and also a small subset of all possible values (including mutants or variants not presently found in the species, but genetically and developmentally possible). For whatever reasons (see section 8.1.3), the range of possible values of X_1 and X_2 is constrained to be within certain limits or zones (marked C_a and C_b); for all traits, these zones represent the "bauplan" of the species, and perhaps even of a genus ("bauplans" of higher taxa such as genera and families would presumably be larger). As a result of selection and/or genetic correlations between the traits, there is no reason to expect that the axes of the population and species ellipses and the constrained limits will *necessarily* be parallel and equal. Of course it is possible that all axes are parallel because natural selection can affect the constraints themselves, but these relationships have not been investigated.

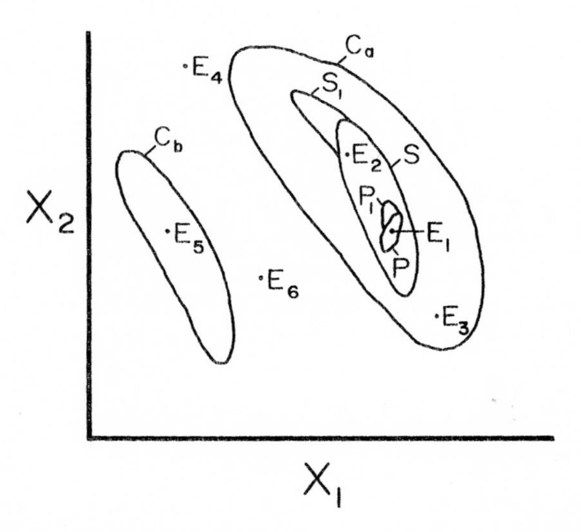

FIGURE 8.1. Population and species variation in reference to constraints. X_1 and X_2 axes represent the range of values of traits 1 and 2. A population is represented as an ellipse P and it can move by mutation and frequency change to P_1, or even further. The range of existing variants in a species is represented by a larger ellipse S, and a species can also expand into S_1 or further. The range of all genetically and developmentally possible values of X_1 and X_2 is constrained to be within certain limits, in this example forming two zones of constrained values C_a and C_b; this is the "bauplan" of the species, if not the genus. The ideal equilibrium combination of trait values in a given environment can lie anywhere in (E_1, E_2, E_3, E_5) or outside (E_4, E_6) the constrained limits. See text for further details.

The "selection" and "random" views implicitly assume that the limits (C) are very large relative to the species ellipse (S), and that there are no disjunct areas within the limits (Wright's shifting balance and Mayr's genetic revolution ideas may be exceptions to the nondisjunction assumption). The "selection" view assumes that most changes in mean position (substitution and replacement) within the limits are due to natural selection, while the "random" view assumes that most are due to genetic drift. In both views, mutation is required to expand the margins of the present population (P) or species (S) ellipses to new territory (P_1, S_1), but the reasons for the actual move (frequency change) are selection or drift, respectively. Note that gene flow can also expand P into P_1, and, similarly, hybridization can expand S into S_1. The "equilibrium" view is similar to the previous two, but in addition assumes that the population or species is at or near an equilibrium position (E_1). Since the position of the equilibrium E will usually depend upon the environment as well as on genetics and development, E may vary geographically and temporally, so there may be more than one equilibrium point for a species. In fact the distribution of E may to some extent determine S, but this has so far not been investigated.

The "constraint" school emphasizes the fact that the genetically and developmentally possible values of X_1 and X_2 may be limited and that major genetic or developmental changes may be necessary to obtain a really new phenotype. In terms of Figure 8.1, the constrained limits (C) are small relative to the species ellipse, and the limits consist of disjunct zones (C_a and C_b). The optimum value (E) may be physically (genetically or developmentally) impossible, or outside the constraints (E_4), or even within a disjunct constraint zone (E_5). In that case it may require a "genetic revolution" or some genetic or developmental reorganization to reach the new zone.

The position of E relative to the population determines the observed mode of selection. If E corresponds to or is close to the mean of a population (as for E_1), then the population experiences stabilizing selection. If E is away from the population mean but inside S and the same constrained zone (E_2), then the population goes through three stages: (1) directional selection until it moves to overlap E_2: (2) both directional and stabilizing selection as it approaches and partially overlaps E_2; and finally (3) only stabilizing selection when its mean corresponds to E_2 (this is equivalent to Figure 1.4*A*, *C*, and *B*). If E is outside the species limits but still in the same constrained limit zone (E_3), mutation is required to reach E_3, whereas only gene flow may be necessary to reach E_2. If E is at or outside the constrained limits (E_4) the population will never reach its predicted optimum value without a genetic or developmental "revolution"; in a practical sense the optimum does not exist and a real plateau will be reached. If E is within a disjunct constrained zone (E_5), a minor genetic and developmental reorganization may be necessary to reach the optimum. If the optimum is outside all constrained zones (E_6), then major genetic and developmental reorganization is required; it must be great enough to change the shape and size of the constrained zones before the population can reach this point. The "constraint" view emphasizes movement *between small disjunct* zones and also emphasizes the origin and determination of the forms of the zones themselves. The other views place more emphasis on movement *within large nondisjunct* zones. Clearly both could be right for different species, and, as mentioned earlier, Atchley's (1984) experiments suggest that the zones themselves are subject to natural selection and genetic drift. The form and evolution of constrained zones would repay further study.

As Mayr (1962, 1978) has long emphasized, evolution is a two-step process, and it involves the *origin* as well as the *replacement* of new variants. The two steps are essential to evolution, and it is myopic if not positively misleading to say that one is more important than another. It is most unfortunate that the two steps have become so separate in people's minds that they tend to be exclusively studied by one or nearly ignored by another group of researchers. The two steps have almost become the provinces of separate fields, as in evolutionary morphology and population genetics. It is high time that this changed; evolutionary studies would be considerably enriched by equal consideration of both origins and replacements.

In order to understand the evolutionary process, it is imperative that we know enough about ecology to predict the course of natural selection and replacement, and enough about genetics and development to understand the origin of new variants. Thorough knowledge can go beyond description and post-facto explanation; it should allow us to predict at least some aspects of evolution. It is extremely difficult if not impossible for a single person to do all of these things, so the temptation to study only one aspect is great. The best thing would be to have a massive increase in collaboration on the same species among ecologists, organismal biologists, developmental biologists, and geneticists rather than to continue the allopatric divergence and speciation of subfields and study organisms that are all too common at the moment. Collaborative work should be done on as many different kinds of animals and plants as possible, so that we do not achieve a biased perception of natural selection and evolution.

8.3. CONCLUSION: NATURAL SELECTION AND EVOLUTION

We have now cleared away enough of the underbrush to have an idea of the importance of natural selection in evolution. Natural selection is common enough in natural populations to have been detected in a wide variety of organisms, and strong selection is not as rare as has been previously assumed; natural selection is therefore likely to be important in evolution. However, natural selection does not explain the origin of new variants, only the process of changes in their frequency. This is the main reason that I have chosen the definitions of the processes of evolution and natural selection given in Chapter 1. The developmental and genetic systems must be known well enough to explain the shape of the constrained limits (*C* in Figure 8.1) and how a species initially appears in a new phenotypic zone. Natural selection (or genetic drift) can only then explain how the species moves through and expands in a zone.

There is only one way in which an understanding of natural selection might help to explain the origins of new traits or trait values, and that is the case where an evolutionary change requires one or more intermediate steps. For example, a given morphological change may require two genetically different adjustments in the morphogenic "program," or the modification of two different enzymes or regulatory pathways in the developmental genetic network. For brevity, call these two mutations (in the general sense) *A* and *B*. Most mutations are rare events, so the probability that both *A* and *B* occur in the same individual or family could easily be as low as 10^{-10}. If the frequency of the new *A* did not change subsequently, and mutation at both "loci" continued, the probability of forming an individual with both *A* and *B* would remain quite small. On the other hand, if *A* were to occur and increase to a high frequency as a result of natural selection, the probability that *B* appears in at least one *A* individual is greatly increased. Consider a third variant *C* which would work as well with *B* as does *A*, but which is not favored by natural selection unless it is present with *B*. It is clear that the combination *AB* would be more likely or more common than *CB*. The effect would be even greater if *B* were functionally impossible unless *A* or *C* occurred first. Thus natural selection may affect the patterns of the origins of combinations of traits, even though it will not explain the mechanisms of their origins. This was tangentially discussed by Fisher (1930), Simpson (1944), and Rensch (1959), but has received virtually no attention since then. It would repay further study.

IMPORTANCE OF SELECTION

There are six major gaps in our knowlege and understanding of natural selection: (1) Why does natural selection occur? What are the biological reasons for the process, and what conditions favor natural selection? (2) How does it occur? What are the mechanisms of natural selection? What is the form of the separation line or selection function? (3) What kinds of traits are most likely to be affected by natural selection? (4) What is the effect of simultaneous natural selection on many traits, some of them intercorrelated with each other? What is the effect of genetic interactions among traits? What is the effect of phenotypic (selective) interactions among traits? Is there any limit to the number of traits that affect fitness, and does this vary with habitat? (5) Given that there is known fitness variation, what are the evolutionary dynamics and equilibrium configurations (if any) of the traits? (6) Is there a relationship between the presence of demonstrable natural selection and genera that are currently radiating rapidly?

The first three questions require extensive studies of the ecology of the species, as well as their predators, parasites, and other factors affecting their life histories. The fourth requires extensive additional data on the genetic system. In order to answer the fifth with any reasonable degree of accuracy, we must have data on the first four. Although there is a massive body of theory for predicting evolutionary dynamics (population and quantitative genetics theory), there is little available data on the models' parameters. All of the questions, especially the sixth, are virtually unexplored. It is tragic that so many studies of natural selection have made little or no attempt to answer the ecological and organismal questions. This is primarily because such an understanding requires a major amount of field work, and also because many researchers have been content with only a superficial impression of their systems, or one that is sufficient merely to demonstrate selection. A lack of action on the fourth question is more excusable, because a general acceptance and realization of its importance, as well as significant advances in methodology, have only recently occurred. The time for superficial ecological genetic studies is over; we need to know the how and why of natural selection.

8.4. SUMMARY

One's opinion on the importance of natural selection depends upon one's relative interests in a variety of aspects of evolution. Different viewpoints emphasize the relative importance of natural selection, mutation, genetic drift, and constraints to variation, as well as their influence on the direction and rate of evolution. The fundamental mechanisms of evolution are the molecular mechanisms leading to new genetic variants, the expression of those variants through the genetic and developmental systems, and constraints to the appearance and function of those variants. Natural selection and genetic drift are mechanisms that cause only frequency changes in populations. Certainly all components are important in evolution, but for different reasons. The tempo of evolution may be most strongly influenced by mutation and genetic drift, and the mode of evolution by natural selection within the constraints of the developmental and genetic systems, which themselves are subject to the same processes. Which aspect of evolution one studies reflects one's interests, and not its relative importance, just as a student of geomorphology must consider erosion, orogeny, and geological structure and history, although he may choose to study only a subset of those phenomena. Increased collaboration among ecologists, geneticists, organismal and developmental biologists is required to cover all aspects of evolution in the same organisms. We are most remarkably ignorant about the ecological aspects of natural selection, and there are at least six major gaps in our knowlege: (1) Why does natural selection occur? (2) How does it occur? (3) What kinds of traits are most likely to be affected? (4) What are the effects of simultaneous natural selection of many traits and of the interactions among them? (5) What are the evolutionary dynamics of selected traits? (6) Are genera that are most prone to exhibit natural selection also those that are currently radiating most rapidly?

> It frequently happens that, by indulging early the raptures of success, we forget the measures necessary to secure it, and suffer the imagination to riot in the fruition of some possible good, till the time of obtaining it has slipped away.
>
> Samuel Johnson, *The Rambler*, 24 March 1750

APPENDIX ONE

Multiple Regression and the Estimation of Selection Differentials

If he is content to take his information from others, he may get through his book with little trouble, and without much endangering his reputation. But if he makes experiments for so comprehensive a book as his, there would be no end to them; his erroneous assertions would then fall upon himself; and he might be blamed for not having made experiments as to every particular.

Samuel Johnson, 1753 (Boswell and Glover 1901, vol. 1, pp. 51–52)

We wish to estimate the effects of directional and variance selection. Before selection, the trait has a mean value $\bar{X}$ and a variance v. We want the estimates of directional and variance selection to be independent of the effects of each other; therefore, following Lande and Arnold (1983), we obtain these estimates from a multiple regression. Let $X_1 = X$, and $X_2 = (X - \bar{X})^2$, and $Y = w$ (relative fitness). The multiple regression equation of relative fitness on the trait value will thus be

$$Y = a + b_1 X_1 + b_2 X_2. \qquad \text{(A.1)}$$

In this equation, b_1 estimates the effects of directional selection on fitness, and b_2 estimates the effects of stabilizing or disruptive selection. This assumes that the trait is phenotypically uncorrelated with other traits. Equation A.1 can be generalized for any number of traits, yielding two regression coefficients b_{1k} and b_{2k} for each trait k. These estimate the effect of phenotypic selection on trait k, holding the effects of the correlated traits constant (see Chapter 6, section 6.2.3, and Lande and Arnold 1983). This appendix will be concerned only with phenotypic selection on a trait that is assumed to be phenotypically uncorrelated with other traits, although section A.1.1 gives the mechanics of multiple regression for any number of intercorrelated traits.

A1.1. MECHANICS OF MULTIPLE REGRESSION

Snedecor and Cochran (1967) give an unusually clear description of how multiple regression works, and this is the basis for the regression formulae in this section. (Readers who are familiar with multiple regression may wish to skip to the next section.) With modification, this summary of multiple regression applies also to the analysis of phenotypic selection of many different traits (see Snedecor and Cochran 1967 for details). Given raw data in the form of Y, X_1 and X_2 measured for each of N individuals, the "corrected" sums of squares are

$$\begin{aligned}
\sum x_k^2 &= \sum X_k^2 - \frac{(\sum X_k)^2}{N} \\
\sum y^2 &= \sum Y^2 - \frac{(\sum Y)^2}{N} \\
\sum x_k y &= \sum X_k Y - \frac{(\sum X_k)(\sum Y)}{N} \\
\sum x_1 x_2 &= \sum X_1 X_2 - \frac{(\sum X_1)(\sum X_2)}{N}.
\end{aligned} \tag{A.2}$$

(The subscript k indicates that the sum is either for $X_1 (k = 1)$ or X_2 $(k = 2)$, to yield $\sum x_1^2$, $\sum x_2^2$, $\sum x_1 y$, and $\sum x_2 y$). Note that the parametric covariances are $\mathrm{COV}[X_k, Y] = (\sum x_k y)/N$, and $\mathrm{COV}[X_1, X_2] = (\sum x_1 x_2)/N$.

The determinant of the variance-covariance matrix is

$$D = (\sum x_1^2)(\sum x_2^2) - (\sum x_1 x_2)^2. \tag{A.3}$$

The two regression coefficients are

$$b_1 = \frac{(\sum x_2^2)(\sum x_1 y) - (\sum x_1 x_2)(\sum x_2 y)}{D} \qquad \text{(A.4a)}$$

$$b_2 = \frac{(\sum x_1^2)(\sum x_2 y) - (\sum x_1 x_2)(\sum x_1 y)}{D}, \qquad \text{(A.4b)}$$

and the intercept is

$$a = \bar{Y} - b_1 \bar{X}_1 - b_2 \bar{X}_2. \qquad \text{(A.5)}$$

The standardized regression coefficients are

$$b'_k = b_k \sqrt{\frac{\sum x_k^2}{\sum y^2}}. \qquad \text{(A.6)}$$

If the assumptions of multiple regression hold, then we can estimate the standard error of the b_k, and also test the b_k and the overall regression (equation A.1) for significance. We obtain this as follows: The explained sum of squares is

$$\sum \hat{y}^2 = b_1 \sum x_1 y + b_2 \sum x_2 y \qquad \text{(A.7)}$$

and the residual sum of squares is

$$\sum d^2 = \sum y^2 - \sum \hat{y}^2. \qquad \text{(A.8)}$$

Therefore the overall test for significance is

$$F_{\{2,N-3\}} = \frac{(N-3)\sum \hat{y}^2}{2\sum d^2}. \qquad \text{(A.9)}$$

The standard errors of the regression coefficients, b_k, are

$$S_{b_1} = \sqrt{\frac{\sum d^2 \sum x_2^2}{D(N-3)}} \qquad \text{(A.10a)}$$

$$S_{b_2} = \sqrt{\frac{\sum d^2 \sum x_1^2}{D(N-3)}}. \qquad \text{(A.10b)}$$

From this we obtain the tests for the significance of the b_k:

$$t_{k\{N-3\}} = \frac{b_k}{S_{b_k}}. \qquad \text{(A.11)}$$

It is easy to generalize these equations for more than two X_k (Snedecor and Cochran 1967).

A1.2. MULTIPLE REGRESSION OF Y ON X AND $(X - \bar{X})^2$

In order to convert equations A.2-A.11 into equations of X and $(X - \bar{X})^2$, it is useful to consider moments about the mean. The kth moment about the mean is

$$m_k = \frac{1}{N} \sum (X - \bar{X})^k. \tag{A.12}$$

Using equation A.12 in A.2, we obtain

$$\sum x_1^2 = Nm_2 \qquad \sum x_2^2 = N(m_4 - m_2^2)$$
$$\sum x_1 x_2 = Nm_3 \tag{A.13}$$

with the remaining expressions in equations A.2 unchanged. Using equations A.13, the determinant D (equation A.3) simplifies to

$$D = N^2[m_2(m_4 - m_2^2) - m_3^2], \tag{A.14}$$

and the regression coefficients become

$$b_1 = \frac{(m_4 - m_2^2)\,\mathrm{COV}[X, Y] - m_3\,\mathrm{COV}[(X - \bar{X})^2, Y]}{m_2(m_4 - m_2^2) - m_3^2} \tag{A.15a}$$

$$b_2 = \frac{m_2\,\mathrm{COV}[(X - \bar{X})^2, Y] - m_3\,\mathrm{COV}[X, Y]}{m_2(m_4 - m_2^2) - m_3^2} \tag{A.15b}$$

If X follows the normal distribution, then $m_2 = v$ (the variance), $m_3 = 0$, and $m_4 = 3v^2$. This results in considerable simplification equation of A.15:

$$b_1 = \frac{\mathrm{COV}[X, Y]}{v} \tag{A-16a}$$

$$b_2 = \frac{\mathrm{COV}[(X - \bar{X})^2, Y]}{2v^2}. \tag{A.16b}$$

These are the same as the univariate regressions of Y on X_1 or X_2, for, when X is normally distributed, $\mathrm{COV}[X, (X - \bar{X})^2] = 0$ (no skewness). Note that Lande and Arnold (1983, equation 16) removed the factor $\frac{1}{2}$ from the variance selection coefficient b_2 (equation A.16b). This was in order to equate b_2 with the curvature (second derivative) of the relative fitness function (or surface for more than two traits), enabling the use of the coefficients in predicting the genetic response to selection.

A1.3. RELATIONSHIP BETWEEN SELECTION DIFFERENTIALS AND REGRESSION COEFFICIENTS

The preceding two sections dealt with the estimation of regression coefficients directly from raw data. Sometimes it may be interesting or useful to express the coefficients in terms of changes in the mean and variance of a population before and after selection; these are the selection differentials (equations 6.1 and 6.7). Let $\bar{X}_a$, v_a, and N_a be the mean, variance, and numbers after selection, respectively (as for example, in cohort analysis). If $f(X)$ is the distribution of X before selection $[\sum f(X) = N]$ and $W(X)$ is the absolute fitness of individuals with trait value X, then the mean absolute fitness is $\bar{W} = \sum f(X)W(X)/N$, and the number of individuals after selection is $N_a = \sum f(X)W(X)$. For prediction of trait frequency distributions (Lande and Arnold 1983; Arnold and Wade 1984a,b), we actually want the distribution to have a mean fitness equal to 1, so we convert to relative fitness by $w(X) = W(X)/\bar{W}$ for all X. Associated with the relative fitness distribution is the variance in relative fitness, which we shall call P:

$$P = \frac{\sum(w(X))^2 f(X)}{(N-1)} - \frac{\{\sum w(X) f(X)\}^2}{N(N-1)}. \qquad \text{(A.17)}$$

Note, therefore, that in equations A.2 and following equations, $\sum y^2 = (N-1)P$.

Given the distribution of relative fitness $w(X)$, we obtain the mean ($\bar{X}_a$) and variance (v_a) after selection:

$$\bar{X}_a = \frac{\sum X w(X) f(X)}{N_a} \tag{A.18a}$$

$$v_a = \frac{\sum X^2 w(X) f(X)}{(N_a - 1)} - \frac{\{\sum X w(X) f(X)\}^2}{N_a(N_a - 1)}. \tag{A.18b}$$

As in equation A.17, these equations are in terms of the observed estimates of $\bar{X}_a$ and v_a, rather than the parametric values.

Substituting the relationship between the old and new distributions of X into equation A.2, after some algebraic manipulation the covariances in equation A.15 and A.16 become

$$\mathrm{COV}[X_1, Y] = \mathrm{COV}[X, w] = \bar{X}_a - \bar{X} \tag{A.19a}$$

$$\mathrm{COV}[X_2, Y] = \mathrm{COV}[(X - \bar{X})^2, w] = v_a - v + (\bar{X}_a - \bar{X})^2. \tag{A.19b}$$

Given the assumption of a normal distribution of X, and substituting equation A.19 in A.16, the regression coefficients become

$$b_1 = \frac{\bar{X}_a - \bar{X}}{v} \tag{A.20a}$$

$$b_2 = \frac{v_a - v + (\bar{X}_a - \bar{X})^2}{2v^2} \tag{A.20b}$$

From equation A.6, the standardized regression coefficients are

$$b'_1 = \frac{\bar{X}_a - \bar{X}}{\sqrt{vP}} = \frac{i}{\sqrt{P}} \tag{A.21a}$$

$$b'_2 = \frac{v_a - v + (\bar{X}_a - \bar{X})^2}{v\sqrt{2P}} = \frac{j + i^2}{\sqrt{2P}}, \tag{A.21b}$$

where $i = (\bar{X}_a - \bar{X})/\sqrt{v}$ and $j = (v_a - v)/v$, from equations 6.1

and 6.2 in Chapter 6. The term i^2 corrects for the reduction in variance due to directional selection, which is one of the aims of multiple regression. It will be an incomplete correction if X is not normally distributed (equation A.15). Note once again that, to use b_k b'_k in predicting the response to selection, one must remove the factor $1/2$ or $1/\sqrt{2}$ (respectively). This is the derivation of the corrected standardized selection differentials i^* and j^* (equation 6.7) from the standardized multiple regression coefficients. Corrected standardized selection differentials can be derived for each of many phenotypically intercorrelated traits in a similar manner, using equation A.6.

A1.4. ALTERNATIVE DERIVATION OF EQUATIONS A.20 AND A.21

Section A1.3 derived equations A.20 and A.21 assuming a continuous distribution of fitnesses. But in many cases available data will only allow division into two classes: survived or died, mated or unmated, laid eggs or not, set seed or not, and so forth. This special case makes the algebra much simpler. Let the survived (mated, laid, set, etc.) class a have a fitness of $W_a = 1.0$ and the other class b have a fitness of $W_b = 0.0$. Calculate the mean absolute fitness $\bar{W}$, and convert the two fitnesses to relative fitness: $w_a = 1/\bar{W}$ and $w_b = 0$. Now $\sum w = N$, $\bar{w} = 1.0$, and $\sum y^2 = N[(1 - \bar{W})/\bar{W}]$. Therefore the estimated variance of relative fitness is:

$$P = \frac{N}{(N-1)} \frac{(1-\bar{W})}{\bar{W}}. \tag{A.22}$$

Note that the number of individuals in class a (the number after selection) is $N\bar{W}$; call this N_a. Since $w_b = 0 = Y_b$, the quantity $\sum X_1 Y$ reduces to $(1/\bar{W})\sum X_a$ (where $\sum X_a$ is for group a), and similarly for $\sum X_2 Y$. With very little algebra this leads to $\mathrm{COV}[X_1, Y] = \mathrm{COV}[X, w] = \bar{X}_a - \bar{X}$, and $\mathrm{COV}[X_2, Y] = \mathrm{COV}[(X - \bar{X})^2, w] = (v_a - v) + (\bar{X}_a - \bar{X})^2$, as in equation

A.20. Equations A.21 simplify further to:

$$b_1' = \frac{i}{\sqrt{P}} = i\sqrt{\frac{\bar{W}}{(1-\bar{W})}\,\frac{(N-1)}{N}} \qquad \text{(A.23a)}$$

$$b_2' = \frac{j+i^2}{\sqrt{2P}} = (j+i^2)\sqrt{\frac{\bar{W}}{(1-\bar{W})}\,\frac{(N-1)}{2N}}\,. \qquad \text{(A.23b)}$$

I recommend the alternative derivation as an exercise to anyone who is unfamiliar with multiple regression and unused to thinking about how fitness affects trait frequency distributions.

A1.5. REGRESSION USING X^2 INSTEAD OF $(X-\bar{X})^2$

Lande and Arnold (1983) suggest using X^2 as X_2 in the multiple regression to estimate variance selection. Their statement actually applies to the deviations from the mean ($x = X - \bar{X}$) rather than raw X values, though this is not obvious from a first reading. It is actually possible to use $X_2 = X^2$ in the multiple regression if the final calculated values are modified as described below.

Expanding equation A.1 in terms of $X_1 = X$ and $X_2 = (X - \bar{X})^2$:

$$Y = (a + b_2\bar{X}^2) + (b_1 - 2\bar{X}b_2)X + b_2X^2, \qquad \text{(A.24)}$$

we would *observe* the following by doing a multiple regression of Y on X and $\bar{X}^2$:

$$Y = A + B_1X + B_2X^2. \qquad \text{(A.25)}$$

In this case B_2 would estimate b_2 [the relationship between Y and $(X-\bar{X})^2$], but B_1 would *not* estimate b_1 [the linear relationship between Y and X]. In this case, b_1 must be estimated by

$$b_1 = B_1 + 2\bar{X}B_2. \qquad \text{(A.26)}$$

If this correction were not made, the effects of directional selection would be underestimated. This caution does not apply to the direct use of equations A.20 or A.21, though they do require the assumption of normality. The reason for the correction is that, for "pure" variance selection, when the mean trait value does *not* correspond to the trait value that has the highest fitness, there will be a directional selection coefficient. On the other hand, there is no directional selection component when X is at the optimum, and both the fitness function and the phenotypic distribution are symmetrical.

APPENDIX TWO

Comparisons Between Selection Differentials and Regression Coefficients Using Simulated Data of Selection with Known Properties

> To convince any man against his will is hard, but to please him against his will is justly pronounced by Dryden to be above the reach of human abilities. Interest and passion will hold out long against the closest siege of diagrams and syllogisms.
>
> Samuel Johnson, *The Rambler*, 5 February 1751

Consider individuals that vary in the values of two traits, X_1 and X_2. Three sets of simulated individuals were studied. Their trait distributions were generated at random from a bivariate normal distribution with parametric means of 50 and standard deviations of 10 for both traits, and three parametric correlations 0.0, 0.7 and -0.7 (Chapter 6, Figure 6.2A,B,C). The observed correlations between X_1 and X_2 are: -0.009, 0.693, and -0.664. These distributions were then subject to selection with known geometry.

Selection was imposed on the "populations" by means of threshold, which was a function of the trait values ("truncation" selection) and separates individuals with high and low fitness. Probabilistic selection (a gradient rather than an absolute threshold) yields similar results, but only the threshold model is presented here because it gives the multivariate methods the best possible chance to work. Three forms of selection were imposed:

(1) directional selection; (2) a combination of directional and density-dependent selection; and (3) variance (stabilizing or disruptive) and covariance selection. The threshold or *selection line* divides the population into two groups, and either group can be thought of as the survivors or the nonsurvivors. For each selection model, individuals of one group were assigned an absolute fitness of $W = 1$ and the others a fitness of $W = 0$. The multiple regression, discriminant function, and estimated selection lines were calculated. For the discriminant function analysis, the means and variance-covariance matrices were estimated from the data, and both groups were assumed to share a common covariance matrix (equation 6.20). The estimated selection line will be called the *separation line* for brevity; it is perpendicular to the discriminant function. For comparison between the separation and selection lines, the standard error of the separation line can be estimated. Ideally it should be estimated using the standard error of the d_k, but those errors are poorly known and complex (Lachenbruch 1975). However, since (for two groups) b_k and d_k are proportional (equation 6.29), we can use the standard errors of b_k (equation A.10). Since the slope of the separation line is ratio of two d_k (equation 6.23), which are themselves related to b_k, we can obtain a crude estimate of the error from the formula for the variance of the ratio of two variables with known variances:

$$\mathrm{VAR}(x/y) = \frac{\mathrm{VAR}(x)}{\bar{y}^2} + \frac{\bar{x}^2\,\mathrm{VAR}(y)}{\bar{y}^4} - \frac{2\bar{x}\,\mathrm{COV}(x,y)}{\bar{y}^3}.$$

Since the b_k should not be correlated with each other, we can ignore the last term in the calculation of the variance of b_k/b_l. The selection differentials (i and j) and univariate selection coefficients (i^* and j^*) were also calculated for X_1, X_2, and D. The results are summarized in Tables A.1–A.6 and section 6.4.6 of Chapter 6.

Directional selection (Figure 6.2, Table A.1). The actual selection line was $X_2 = A + BX_1$, where A is the intercept and B

is the slope. Various slopes were chosen so that the line went through the population mean or was raised above the mean of X_2 (Table A.1). When B is at or near zero, then the contribution to fitness of trait X_1 is small; when B is very large, then X_2 contributes little to fitness.

Directional selection with density dependence (Figure 6.3, Tables A.2 and A.3). In addition to directional selection, it is possible for the probability of survival of a particular phenotype to be negatively correlated to its density (or frequency). This will cause a "bump" in the directional selection line. The bump can be defined as a given constant probability contour of the bivariate normal distribution. The probability contour times the total population size defines the critical density above which fitness declines, and this, together with the directional selection lines, was imposed on the population. Two examples of the combined selection lines are shown in Figure 6.3.

Variance and covariance selection (Figure 6.4). The selection line (in this case a circle or ellipse) was determined by a constant probability contour in the bivariate normal distribution.

Variance-covariance matrices of high and low fitness groups. Table A.4 shows the effect of various forms of selection of the variance-covariance matrices. Since there are only two traits with equal variances in this artificial data set, for brevity only the correlations between X_1 and X_2 are shown.

Univariate measures. Tables A.5 and A.6 compare the results of univariate and multivarite measures of selection, as well as the results of univariate measures calculated on the raw data with those calculated on the discriminant function axis (D).

Actual Line[a]		Multiple Regression[b]					Multiple r^2	Estimated Line		Percent Wrong	$\bar{W}^c$
Slope	Incpt.	b_1	b_2	Incpt.	b_2'	b_2'		Slope	Incpt.		
				parametric $r = 0.0$							
−1.0	100	*0.0274*	*0.0331*	−2.5866	0.550	0.581	0.635	−0.827	93.315	2	0.46
		0.0031	0.0035								
−1.0	110	*0.0247*	*0.0257*	−2.2596	0.551	0.501	0.550	−0.961	106.951	4	0.28
		0.0031	0.0035								
−0.5	75	*0.0182*	*0.0417*	−2.5008	0.365	0.731	0.663	−0.436	71.946	2	0.52
		0.0029	0.0034								
0	50	0.0051	*0.0454*	−1.9862	0.103	0.800	0.650	−0.112	54.786	5	0.56
		0.0030	0.0034								
0	60	−0.0022	*0.0225*	−0.9040	−0.068	0.606	0.373	0.098	57.794	4	0.12
		0.0026	0.0030								
0.1	45	−0.0016	*0.0449*	−1.6447	−0.032	0.789	0.623	0.036	47.776	2	0.54
		0.0031	0.0035								
0.5	25	*−0.0233*	*0.0372*	−0.2174	−0.466	0.652	0.648	0.625	19.272	1	0.49
		0.0030	0.0034								
1.0	0	*−0.0322*	*0.0286*	0.6599	−0.647	0.501	0.675	1.127	−5.585	0	0.48
		0.0029	0.0033								
1.0	10	*−0.0226*	*0.0224*	0.2377	−0.539	0.467	0.512	1.009	10.573	1	0.23
		0.0030	0.0034								
10.0	−450	*−0.0393*	*0.0073*	2.1328	−0.788	0.127	0.639	5.420	−225.218	5	0.52
		0.0030	0.0035								

(continued)

TABLE A.1. *(continued)*

Actual Line[a]		Multiple Regression[b]					Multiple	Estimated Line		Percent	
Slope	Incpt.	b_1	b_2	Incpt.	b'_2	b'_2	r^2	Slope	Incpt.	Wrong	$\bar{W}^c$
					parametric $r = 0.7$						
−1.0	100	*0.0169* 0.0048	*0.0289* 0.0040	−1.7663	0.287	0.589	0.663	−0.583	78.423	3	0.52
−1.0	110	*0.0169* 0.0053	*0.0217* 0.0045	−1.6145	0.309	0.474	0.523	−0.780	97.102	3	0.32
−0.5	75	*0.0131* 0.0048	*0.0317* 0.0040	−1.7249	0.223	0.646	0.666	−0.412	70.091	0	0.51
0	50	0.0035 0.0048	*0.0377* 0.0040	−1.5364	0.060	0.768	0.657	−0.093	53.955	1	0.51
0	60	−0.0036 0.0043	*0.0214* 0.0036	−0.7556	−0.095	0.670	0.370	0.170	53.813	4	0.12
0.1	45	0.0016 0.0049	*0.0388* 0.0041	−1.4823	0.028	0.790	0.655	−0.042	51.110	1	0.52
0.5	25	*−0.0265* 0.0049	*0.0514* 0.0041	−0.7108	−0.451	1.046	0.643	0.515	23.575	1	0.49
1.0	0	*−0.0540* 0.0051	*0.0519* 0.0043	0.6395	−0.922	1.059	0.618	1.040	−2.683	3	0.47
1.0	7	*−0.0377* 0.0047	*0.0395* 0.0039	0.1681	−0.789	0.987	0.517	0.955	7.600	3	0.21
1.0	8	*−0.0264* 0.0041	*0.0272* 0.0035	0.1118	−0.692	0.852	0.388	0.970	10.549	5	0.12
1.0	10	*−0.0197* 0.0037	*0.0208* 0.0031	0.0459	−0.617	0.782	0.324	0.944	14.108	1	0.08

					parametric $r = -0.7$						
−1.0	100	*0.0499*	*0.0550*	−4.7424	1.056	0.992	0.708	−0.907	95.375	2	0.50
		0.0035	0.0041								
−1.0	107	*0.0367*	*0.0304*	−3.1516	0.954	0.673	0.510	−1.207	119.113	3	0.15
		0.0037	0.0043								
−1.0	108	*0.0288*	*0.0268*	−2.6335	0.853	0.679	0.419	−1.073	114.172	4	0.21
		0.0035	0.0041								
−1.0	110	*0.0222*	*0.0208*	−2.0535	0.785	0.625	0.355	−1.071	116.814	3	0.10
		0.0031	0.0036								
−0.5	75	*0.0317*	*0.0593*	−4.0346	0.673	1.070	0.641	−0.536	76.502	5	0.50
		0.0038	0.0045								
0	50	−0.0042	*0.0419*	−1.3725	−0.089	0.757	0.670	0.100	44.700	3	0.48
		0.0037	0.0043								
0	60	−0.0017	*0.0195*	−0.7555	−0.054	0.541	0.344	0.085	58.829	6	0.12
		0.0034	0.0040								
0.1	45	*−0.0074*	*0.0393*	−1.0777	−0.158	0.710	0.678	0.190	40.163	1	0.48
		0.0036	0.0043								
0.5	25	*−0.0128*	*0.0341*	−0.5185	−0.272	0.616	0.676	0.376	29.865	1	0.51
		0.0037	0.0043								
1.0	0	*−0.0207*	*0.0252*	0.3064	−0.438	0.455	0.664	0.819	7.676	3	0.50
		0.0037	0.0044								
1.0	10	*−0.0197*	*0.0202*	0.3194	−0.452	0.394	0.596	0.979	8.645	1	0.31
		0.0038	0.0044								
10.0	−450	*−0.0293*	*0.0128*	1.3606	−0.621	0.230	0.628	2.297	−67.478	8	0.50
		0.0039	0.0046								

[a] Slope and intercept given for selection line drawn through parametric mean (50.0, 50.0); actual selection line drawn through observed mean.

[b] b values italicized if significant ($P < 0.05$); numbers below b are standard errors.

[c] Mean fitness assuming individuals above selection line have survived; equal to $1 - \bar{W}$ if individuals below selection line survived.

TABLE A.2. Multiple regression and discriminant function descriptions of directional selection and density-dependence, using distributions of Figure 6.2A,B,C

Actual Line		Multiple Regression					Multiple	Estimated Line[a]		Percent	
Slope	Incpt.	b_1	b_2	Incpt.	b'_1	b'_2	r^2	Slope	Incpt.	Wrong	$\bar{W}$
					Parametric $r = 0.0$						
−1.0	100	*0.0164*	*0.0230*	−1.7580	0.391	0.478	0.379	−0.715	96.809	9	0.23
		0.0034	0.0038								
0	50	0.0011	*0.0314*	−1.3628	0.025	0.613	0.376	−0.036	58.775	16	0.28
		0.0036	0.0041								
1.0	0	*−0.0261*	*0.0201*	0.5818	−0.593	0.391	0.496	1.301	−4.719	11	0.28
		0.0032	0.0037								
					Parametric $r = 0.7$						
−1.0	100	0.0051	*0.0188*	−1.0008	0.110	0.489	0.326	0.270	76.746	8	0.19
		0.0053	0.0045								
0	50	−0.0049	*0.0356*	−0.8246	−0.085	0.635	0.434	0.161	51.561	8	0.20
		0.0065	0.0046								
1.0	0	*−0.0374*	*0.0331*	0.4609	−0.797	0.842	0.415	1.131	−0.122	10	0.20
		0.0051	0.0042								
					Parametric $r = -0.7$						
−1.0	100	*0.0365*	*0.0336*	−3.2733	0.905	0.711	0.470	−1.085	111.444	8	0.24
		0.0040	0.0047								
0	50	*−0.0090*	*0.0215*	−0.3948	−0.234	0.478	0.431	0.417	39.797	5	0.21
		0.0039	0.0046								
1.0	0	*−0.0213*	0.0088	0.8971	−0.521	0.184	0.433	2.410	−47.473	10	0.25
		0.0042	0.0049								

TABLE A.3. Multiple quadratic regression of data used for Table A.2

	Slope = 0.0		Slope = −1.0		Slope = 1.0	
	b	Standard Error	b	Standard Error	b	Standard Error
			Parametric $r = 0.0$			
X_1	*−0.1050*	0.0240	*−0.0862*	0.0218	*−0.1394*	0.0185
X_2	*−0.0827*	0.0297	*−0.0979*	0.0269	*−0.1306*	0.0228
X_1^2	*0.0011*	0.0002	*0.0010*	0.0002	*0.0011*	0.0002
X_2^2	*0.0012*	0.0003	*0.0012*	0.0003	*0.0015*	0.0002
A	3.8708		3.5550		6.8657	
R^2	0.376		0.379		0.728	
			Parametric $r = 0.7$			
X_1	*−0.0762*	0.0380	*−0.0746*	0.0361	*−0.0724*	0.0369
X_2	−0.0395	0.0254	*−0.0513*	0.0242	−0.0210	0.0247
X_1^2	*0.0008*	0.0004	*0.0008*	0.0004	0.0004	0.0004
X_2^2	*0.0006*	0.0003	*0.0007*	0.0002	*0.0005*	0.0002
A	2.3901		2.5243		2.5506	
R^2	0.444		0.478		0.475	
			Parametric $r = -0.7$			
X_1	*−0.0977*	0.0182	*−0.0542*	0.0204	*−0.1107*	0.0213
X_2	*−0.0783*	0.0215	−0.0212	0.0242	*−0.0610*	0.0253
X_1^2	*0.0009*	0.0002	*0.0009*	0.0002	*0.0009*	0.0002
X_2^2	*0.0010*	0.0002	*0.0006*	0.0002	*0.0007*	0.0003
A	4.0664		0.2021		4.6872	
R^2	0.658		0.607		0.585	

NOTE: b values italicized if significant at least $P < 0.05$.

TABLE A.4. Correlation between X_1 and X_2 in groups separated by a selection line or ellipse

Actual Line		Population Correlation and Group Number					
		−0.009		0.693		−0.664	
Slope	Incpt.	1	2	1	2	1	2
		Directional (Linear) Selection Line (Figure 6.2)					
−1.0	100	−0.440	−0.517	0.361	0.290	−0.902	−0.873
−1.0	100	−0.316	−0.668	0.498	0.256	−0.737	−0.963
−0.5	75	−0.430	−0.411	0.390	0.303	−0.823	−0.817
0	50	−0.138	−0.147	0.492	0.402	−0.374	−0.414
0	60	0.029	0.413	0.653	0.427	−0.600	0.022
0.1	45	0.071	0.001	0.509	0.434	−0.335	−0.340
0.5	25	0.444	0.462	0.776	0.725	−0.289	−0.262
1.0	0	0.367	0.602	0.874	0.856	−0.254	−0.246
1.0	10	0.251	0.747	0.749	0.997	−0.407	−0.031
10.0	−450	0.087	0.221	0.536	0.615	−0.311	−0.386
		Directional Selection and Density-Dependence (Figure 6.3)					
0	50	0.028	−0.156	0.686	0.468	−0.588	0.017
−1.0	0	−0.093	−0.693	0.658	0.191	−0.727	−0.920
1.0	0	0.027	0.734	0.755	0.920	−0.608	−0.093
		Variance and Covariance Selection (Figure 6.4)					
0.0[a]		−0.045	0.186	0.755	0.414		
0.0		0.048	−0.066	0.856	0.452		
0.7		−0.068	0.589	0.692	0.644		
0.7		−0.152	0.383				
−0.7				0.764	−0.399		
−0.7				0.816	−0.147		

[a] Correlation of selection threshold surrounding group 2.

TABLE A.5. Comparative performance of univariate and multivariate measures of directional selection: Results of significance tests for i, j, and b_k

	$r = 0.0$		$r = 0.7$ or -0.7	
	b_k Not Sig.	b_k Sig.	b_k Not Sig.	b_k Sig.
	Directional Selection			
i not significant	3	1	4	8
i significant	0	16	6	30
j significant	19	1	40	0
j significant	0	0	8	0
	Directional and Density-dependent Selection			
i not significant	1	0	0	1
i significant	0	5	3	8
j not significant	0	4	1	4
j significant	0	2	1	6
	Variance and Covariance Selection			
Not significant	(i) 18	0	(j) 0	0
Significant	(i) 0	0	(j) 0	18

TABLE A.6. Comparison of i, i', j, j' when there is some selection on the variance

Actual line		Parametric		i or j[a]			i' or j'		
Slope	Incpt.	r		X_1	X_2	D	X_1	X_2	D
Directional Selection and Density-Dependence									
−1.0	100	0.0	i	*0.705*	*0.865*	*1.120*	0.383	0.470	0.6
			j	0.030	−0.094	*−0.647*	0.287	0.356	0.3
0	50	0.0	i	0.032	*0.977*	*0.978*	0.020	0.606	0.6
			j	*0.527*	−0.530	−0.539	0.328	0.263	0.2
1.0	0	0.0	i	*−0.935*	*0.632*	*1.124*	−0.580	0.392	0.6
			j	−0.125	*0.272*	*−0.721*	0.465	0.417	0.3
−1.0	100	0.7	i	*−0.922*	*1.161*	*1.172*	0.444	0.550	0.5
			j	−0.319	−0.441	−0.526	0.256	0.437	0.4
0	50	0.7	i	*0.692*	*1.125*	*1.132*	0.344	0.560	0.5
			j	0.062	−0.455	−0.437	0.269	0.403	0.4
1.0	0	0.7	i	*−0.425*	*0.577*	*1.281*	−0.211	0.287	0.6
			j	*0.617*	*0.475*	−0.602	0.397	0.402	0.5
−1.0	100	−0.7	i	*0.767*	0.194	*1.214*	0.429	0.108	0.6
			j	*0.255*	*0.468*	*−0.651*	0.472	0.283	0.4
0.0	50	−0.7	i	*−1.063*	*1.221*	*1.267*	−0.545	0.626	0.6
			j	−0.199	*−0.723*	*−0.756*	0.478	0.394	0.4
1.0	0	−0.7	i	*−1.109*	*0.914*	*1.134*	−0.637	0.525	0.6
			j	−0.512	−0.326	−0.618	0.412	0.293	0.3

BLE A.6. (*continued*)

Selection		Param.		*i* or *j*			*i'* or *j'*[b]		
r	$\bar{W}$	*r*		X_1	X_2	*D*	X_1	X_2	*D*
				Stabilizing Selection					
.0	0.50	0.0	*i*	0.066	0.006	0.066	0.066	0.006	0.066
			j	*−0.695*	*−0.674*	*−0.684*	−0.687	−0.671	−0.676
.0	0.78	0.0	*i*	−0.014	0.071	0.072	−0.026	0.133	0.135
			j	*−0.417*	*−0.401*	*−0.389*	−0.781	−0.742	−0.719
.7	0.35	0.0	*i*	−0.022	−0.050	0.055	−0.016	−0.037	0.040
			j	*−0.766*	*−0.707*	*−0.600*	−0.559	−0.514	−0.436
.7	0.60	0.0	*i*	0.061	0.006	0.061	0.074	0.007	0.074
			j	*−0.535*	*−0.572*	*−0.495*	−0.647	−0.697	−0.599
0	0.59	0.7	*i*	0.085	0.067	0.086	0.101	0.080	0.103
			j	*−0.674*	*−0.685*	*−0.704*	−0.796	−0.812	−0.831
.0	0.82	0.7	*i*	*0.122*	0.090	*0.123*	0.259	0.191	0.261
			j	*−0.500*	*−0.481*	*−0.519*	−1.030	−1.004	−1.070
.7	0.53	0.7	*i*	0.170	0.159	0.179	0.180	0.168	0.189
			j	*−0.646*	*−0.691*	*−0.674*	−0.652	−0.703	−0.678
.7	0.30	0.7	*i*	0.150	−0.053	0.264	0.098	−0.035	0.172
			j	*−0.861*	*−0.761*	−0.207	−0.546	−0.494	−0.089
.7	0.56	0.7	*i*	0.069	0.017	−0.081	0.077	0.019	−0.091
			j	*−0.700*	*−0.643*	*−0.356*	−0.780	−0.721	−0.392

r *j* values italicized if significant.
the quadratic regression $W = a + b_1X_1 + b_2X_2 + b_3X_1^2 + b_4X_2^2$; all quadratic b_k coeffi-
ts were significant, except b_4 for the second to last example ($j' = -0.494$)

References

Let me know whether I have not sent you a pretty library. There are, perhaps, many books among them which you never need to read through; but there are none which it is not proper to know, and sometimes to consult. Of these books, of which the use is only occasional, it is often sufficient to know the contents, that when any question arises, you may know where to look for information.

Samuel Johnson, 1776 (Boswell and Glover 1901, vol 3, p. 283)

Abbott, R. J. 1977. A quantitative association between soil moisture content and the frequency of the cyanogenic form of *Lotus corniculatus* L. at Birsay, Orkney. *Heredity* 38:397–400.

Adams, J., and R. H. Ward. 1973. Admixture studies and the detection of selection. *Science* 180:1137–1143.

Alcock, J. 1979. The evolution of intraspecific diversity in male reproductive strategies in some bees and wasps. Pp. 381–402 in M. S. Blum and N. A. Blum, eds., *Sexual Selection and Reproductive Competition in Insects*. New York: Academic Press.

Alcock, J. 1984. Long-term maintenance of size variation in populations of *Centris pallida* (Hymenoptera: Anthorphoridae). *Evolution* 38:220–223.

Allard, R. W., G. R. Babbel, M. T. Clegg, and A. L. Kahler. 1972. Evidence for coadaptation in *Avena barbata*. *Proc. Nat. Acad. Sci. U.S.A.* 69:3043–3048.

Allard, R. W., and S. K. Jain. 1962. Population studies in predominantly self-pollinated species. II. Analysis of quantitative genetic changes in a bulk-hybrid population of barley. *Evolution* 16:90–101.

Allard, R. W., A. L. Kahler, and M. T. Clegg. 1977. Estimation of mating cycle components of selection in plants. Pp. 1–19 in F. B. Christiansen and T. M. Fenchel, eds., *Measuring Selection in Natural Populations*. New York: Springer-Verlag.

Allard, R. W., A. L. Kahler, and B. S. Weir. 1972. The effect of selection on esterase allozymes in a barley population. *Genetics* 72:489–503.

Allen, W. R., and P. M. Sheppard. 1971. Copper tolerance in some Californian populations of the monkey flower, *Mimulus guttatus*. *Proc. Roy. Soc. Lond. B.* 177:177–196.

Allendorf, F. W., and R. F. Leary. 1986. Heterozygosity and fitness in natural populations of animals. In M. Soulé, ed., *Conservation Biology: The Science of Diversity*. Sunderland, Mass.: Sinauer Associates (in press).

Allison, A. C. 1954. Notes on sickle-cell polymorphisms. *Ann. Hum. Genet.* 19:39–51.

Allison, A. C. 1955. Polymorphism and natural selection in human populations. *Cold Spring Harbor Symp. Quant. Biol.* 29:137–150.

Allison, A. C. 1975. Abnormal haemoglobin and erythrocyte enzyme deficiency traits. Pp. 101–122 in D. F. Roberts, ed., *Human Variation and Natural Selection. Symp. Soc. Study Hum. Biol.*, vol. 13.

Allison, A. C., and D. F. Clyde. 1961. Malaria in African children with deficient erythrocyte glucose-6-phosphate dehydrogenase. *Brit. Med. J.* 1961:1346–1349.

Alstad, D. N., and G. F. Edmunds, Jr. 1983a. Selection, outbreeding depression, and the sex ratio of scale insects. *Science* 220:93–95.

Alstad, D. N., and G. F. Edmunds, Jr. 1983b. Adaptation, host specificity, and gene flow in the Black Pineleaf Scale. Pp. 413–426 in R. F. Denno and M. S. McClure, eds., *Variable Plants and Herbivores*. New York: Academic Press.

Ammerman, A. J., and L. L. Cavalli-Sforza. 1984. *The Neolithic Transition and the Genetics of Populations in Europe*. Princeton, N.J.: Princeton University Press.

Anderson, P. K. 1964. Lethal alleles in *Mus musculus*: Local distribution and evidence for isolation of demes. *Science* 145:177–178.

Anderson, R. M., and R. M. May. 1981. The population dynamics of microparasites and their invertebrate hosts. *Phil. Trans. Roy. Soc. Lond. B.* 291:451–524.

Angseesing, J.P.A. 1974. Selective eating of the acyanogenic form of *Trifolium repens*. *Heredity* 32:73–83.

Angseesing, J.P.A., and W. J. Angseesing. 1973. Field observations on the cyanogenic polymorphism in *Trifolium repens*. *Heredity* 31:276–282.

REFERENCES

Antonovics, J. 1971. The effects of a heterogeneous environment on the genetics of natural populations. *Amer. Sci.* 59:593–599.

Antonovics, J., A. D. Bradshaw, and R. G. Turner. 1971. Heavy metal tolerance in plants. *Adv. Ecol. Res.* 7:1–85.

Arnold, S. J. 1983a. Sexual selection: The interface of theory and empiricism. Pp. 67–107 in P. Bateson, ed., *Mate Choice*. Cambridge, Eng.: Cambridge University Press.

Arnold, S. J. 1983b. Morphology, performance and fitness. *Amer. Zool.* 23:347–361.

Arnold, S. J., and M. J. Wade. 1984a. On the measurement of natural and sexual selection: Theory. *Evolution* 38:709–719.

Arnold, S. J., and M. J. Wade. 1984b. On the measurement of natural and sexual selection: Applications. *Evolution* 38:720–734.

Arnold, S. J., M. J. Wade, and R. Lande. 1986. Measuring selection in natural populations. Manuscript.

Asay, K. H., I. T. Carlson, and C. P. Wilsie. 1968. Genetic variability in forage yield, crude protein percentage, and palatability in reed canarygrass, *Phalaris arundinacea*. *Crop. Sci.* 8:568–571.

Ashton, G. C. 1959. β-Globulin polymorphism and early foetal mortality in cattle. *Nature* 183:404–405.

Ashton, G. C. 1965. Cattle serum transferrins: A balanced polymorphism? *Genetics* 52:983–997.

Aston, J. L., and A. D. Bradshaw. 1966. Evolution in closely adjacent populations. II. *Agrostis stolonifera* in maritime habitats. *Heredity* 21:649–664.

Atchley, W. R. 1984. Ontogeny, timing of development, and genetic variance-covariance structure. *Amer. Nat.* 123:519–540.

Atchley, W. R., and J. J. Rutledge. 1980. Genetic components of size and shape. I. Dynamics of components of phenotypic variability and covariability during ontogeny in the laboratory rat. *Evolution* 34:1161–1173.

Baker, A. J. 1980. Morphometric differentiation in New Zealand populations of the house sparrow (*Passer domesticus*). *Evolution* 34:638–653.

Baker, R. J., R. K. Chesser, B. F. Koop, and R. A. Hoyt. 1983. Adaptive nature of chromosomal rearrangement: Differential fitness in pocket gophers. *Genetica* 61:161–164.

Bantock, C. R. 1974. Experimental evidence for non-visual selection in *Cepaea nemoralis*. *Heredity* 33:409–437.

Bantock, C. R. 1980. Variation in the distribution and fitness of the brown morph of *Cepaea nemoralis* (L). *Biol. J. Linn. Soc. Lond.* 13:47–64.

Bantock, C. R., and J. A. Bayley. 1973. Visual selection for shell size in *Cepaea* (Held.). *J. Anim. Ecol.* 42:247–261.

Bantock, C. R., J. A. Bayley, and P. H. Harvey. 1975. Simultaneous selective predation on two features of a mixed sibling species population. *Evolution* 29:636–649.

Bantock, C. R., and M. Ratsey. 1980. Natural selection in experimental populations of the landsnail *Cepaea nemoralis* L. *Heredity* 44:37–54.

Barber, H. N. 1955. Adaptive gene substitutions in Tasmanian Eucalyptus. I. Genes controlling the development of glaucousness. *Evolution* 9:1–14.

Barber, H. N., and W. D. Jackson. 1957. Natural selection in action in *Eucalyptus*. *Nature* 179:1267–1269.

Barker, J.S.F., and P. D. East. 1980. Evidence for selection following perturbation of allozyme frequencies in a natural population of *Drosophila*. *Nature* 284:166–168.

Barndorff-Nielsen, O. 1977. On conditional inference for deviation from Hardy-Weinberg distribution. Pp. 149–157 in F. B. Christiansen and T. M. Fenchel, eds., *Measuring Selection in Natural Populations*. New York: Springer-Verlag.

Bart, J., and D. S. Robson. 1982. Estimating survivorship when the subjects are visited periodically. *Ecology* 63:1078–1090.

Bateson, P., ed. 1983. *Mate Choice*. Cambridge, Eng.: Cambridge University Press.

Beatson, R. R. 1976. Environmental and genetical correlates of disruptive coloration in the water snake *Natrix s. sipedon*. *Evolution* 30:241–252.

Beecher, W. J. 1951. Convergence in the Coerebidae. *Wilson Bull.* 63:274–287.

Bell, G. 1974. The reduction of morphological variation in natural populations of smooth newt larvae. *J. Anim. Ecol.* 43:115–128.

Bell, G. 1978. Further observations on the fate of morphological variation in a population of smooth newt larvae (*Triturus vulgaris*). *J. Zool. Lond.* 185:511–518.

Bell, M. A. 1976. Reproductive character displacement in threespine sticklebacks. *Evolution* 30:847–859.

Bell, M. A., and T. R. Haglund. 1978. Selective predation of threespine sticklebacks (*Gasterosteus aculeatus*) by garter snakes. *Evolution* 32:304–319.

Bellamy, D., R. J. Berry, M. E. Jakobson, W. Z. Lidicker, Jr., J. Morgan, and H. M. Murphy. 1973. Ageing in an isolated population of the house mouse. *Age and Ageing* 2:235–250.

Bender, E. A., T. J. Case, and M. E. Gilpin. 1984. Perturbation experiments in community ecology: Theory and practice. *Ecology* 65:1–13.

Benson, W. W. 1972. Natural selection for Müllerian mimicry in *Heliconius erato* in Costa Rica. *Science* 176:936–939.

Berry, R. J., W. N. Bonner, and J. Peters. 1979. Natural selection in house mice (*Mus musculus*) from South Georgia (south Atlantic Ocean). *J. Zool. Lond.* 189:385–398.

Berry, R. J., and J. H. Crothers. 1968. Stabilising selection in the dog-whelk (*Nucella lapillus*). *J. Zool. Lond.* 155:5–17.

Berry, R. J., and J. H. Crothers. 1970. Genotypic stability and physiological tolerance in the dog-whelk (*Nucella lapillus*). *J. Zool. Lond.* 162:293–302.

Berry, R. J., and P. E. Davis. 1970. Polymorphism and behaviour in the Arctic Skua (*Stercorarius parasiticus*) (L.). *Proc. Roy. Soc. Lond. B.* 175:255–267.

Berry, R. J., and M. E. Jakobson. 1975. Ecological genetics of an isolated population of the house mouse (*Mus musculus*). *J. Zool. Lond.* 175:523–540.

Berry, R. J., and H. M. Murphy. 1970. The biochemical genetics of an island population of the house mouse. *Proc. Roy. Soc. Lond. B.* 176:87–103.

Berry, R. J., and J. Peters. 1975. Macquarie Island house mice, a genetical isolate on a sub-Antarctic island. *J. Zool. Lond.* 176:375–389.

Berry, R. J., and J. Peters. 1976. Genes, survival, and adjustment in an island population of the house mouse. Pp. 23–48 in S. Karlin and E. Nevo, eds., *Population Genetics and Ecology*. New York: Academic Press.

Berry, R. J., J. Peters, and R. J. Van Aarde. 1978. Subantarctic house mice: Colonization, survival and selection. *J. Zool. Lond.* 184:127–141.

Berven, K. A. 1982. The genetic basis of altitudinal variation in the wood frog *Rana sylvatica*. I. An experimental analysis of life history traits. *Evolution* 36:962–983.

Berven, K. A., D. E. Gill, and S. J. Smith-Gill. 1979. Countergradient selection in the green frog, *Rana clamitans*. *Evolution* 33:609–623.

Bethell, T. 1976. Darwin's mistake. *Harpers Magazine* 252:70–75.

Bishop, J. A. 1969. Changes in genetic constitution of a population of *Sphaeroma rugicauda* (Crustacea: Isopoda). *Evolution* 23:589–601.

Bishop, J. A., L. M. Cook, and J. Muggleton. 1978. The response of two species of moths to industrialization in north-west England. II. Relative fitness of morphs and populations. *Phil. Trans. Roy. Soc. Lond. B.* 281:517–540.

Bishop, J. A., and P. S. Harper. 1970. Melanism in the moth *Gonodontis bidentata*: A cline within the Merseyside conurbation. *Heredity* 25:449–456.

Bishop, J. A., and D. J. Hartley. 1976. The size and age structure of rural populations of *Rattus norvegicus* containing individuals resistant to the anticoagulant poison Warfarin. *J. Anim. Ecol.* 45:623–646.

Bishop, J. A., D. J. Hartley, and G. G. Partridge. 1977. The population dynamics of genetically determined resistance to Warfarin in *Rattus norvegicus* from mid-Wales. *Heredity* 39:389–398.

Bishop, J. A., and M. E. Korn. 1969. Natural selection and cyanogenesis in white clover, *Trifolium repens*. *Heredity* 24:423–430.

Blest, A. D. 1963. Longevity, palatability and natural selection in five species of new world Saturniid moth. *Nature* 197:1183–1186.

Blurton Jones, N. 1978. Natural selection and birthweight. *Ann. Hum. Biol.* 5:487–489.

Boag, P. T., and P. R. Grant. 1981. Intense natural selection in a population of Darwin's finches (Geospizinae) in the Galapagos. *Science* 214:82–85.

Bock, W. J. 1959. Preadaptation and multiple evolutionary pathways. *Evolution* 13:194–211.

Bock, W. J. 1980. The definition and recognition of biological adaptation. *Amer. Zool.* 20:217–227.

Bodmer, W. F. 1968. Demographic approaches to the measurement of differential selection in human populations. *Proc. Nat. Acad. Sci. U.S.A.* 59:690–699.

Bodmer, W. F. 1973. Population studies and the measurement of natural selection, with special reference to the HL-A system. *Israel J. Med. Sci.* 9:1503–1518.

Bookstein, F. L., P. D. Gingerich, and A. G. Kluge. 1978. Hierarchical linear modeling of the tempo and mode of evolution. *Paleobiology* 4:120–134.

Borowsky, R. 1977. Detection of the effects of selection on protein polymorphisms in natural populations by means of a distance analysis. *Evolution* 31:341–346.

Boswell, J., and A. Glover. 1901. *The Life of Samuel Johnson.* 3 vols. London: J. M. Dent & Co.

Box, E. O. 1981. *Tasks for Vegetation Science I: Macroclimate and Plant Forms: An Introduction to Predictive Modelling in Phytogeography.* The Hague: W. Junk.

Bradshaw, A. D., T. S. McNeilly, and R.P.G. Gregory. 1965. Industrialisation, evolution, and the development of heavy metal tolerance in plants. *Ecology and the Industrial Society, Brit. Ecol. Soc. Symp.* 5:327–343.

Brady, R. H. 1979. Natural selection and the criteria by which a theory is judged. *Syst. Zool.* 28:600–621.

Brandon, R. N. 1982. The levels of selection. Pp. 315–322 in P. Asquith and T. Nickles, eds., *P.S.A. 1982*, vol. 1. East Lansing, Mich.: Philosophy of Science Association.

Brandon, R. N., and R. M. Burian. 1984. *Genes, Organisms, Populations: Controversies over the Units of Selection.* Cambridge, Mass.: M.I.T. Press.

Brayton, R. D., and B. Capon. 1980. Palatability, depletion, and natural selection of *Salvia columbiarae* seeds. *Aliso* 9:581–587.

Briscoe, D. A., A. Robertson, and J.-M. Malpica. 1975. Dominance at the ADH locus in response of adult *Drosophila melanogaster* to environmental alcohol. *Nature* 255:148–149.

Brown, A.H.D. 1970. The estimation of Wright's fixation index from genotypic frequencies. *Genetica* 41:399–406.

Brown, A.H.D. 1979. Enzyme polymorphisms in plant populations. *Theor. Pop. Biol.* 15:1–42.

Brown, A.H.D., M. W. Feldman, and E. Nevo. 1980. Multilocus structure of natural populations *Hordeum spontaneum*. *Genetics* 96: 523–536.

Brown, A.H.D., D. R. Marshall, and J. Munday. 1976. Adaptedness of variants at an alcohol dehydrogenase locus in *Bromus mollis* L. (soft bromegrass). *Aust. J. Biol. Sci.* 29:389–396.

Brown, A.W.A. 1978. *Ecology of Pesticides*. New York: Wiley.

Brown, A.W.A., and R. Pal. 1971. *Insecticide Resistance in Arthropods*. Geneva: World Health Organization.

Brown, L. N. 1965. Selection in a population of house mice containing mutant individuals. *J. Mammal.* 46:461–465.

Brown, P. J. 1981. New considerations on the distribution of Malaria, Thalassemia, and Glucose-6-Phosphate Dehydrogenase deficiency in Sardinia. *Human Biology* 53:367–382.

Brown, W. K., and E. O. Wilson. 1956. Character displacement. *Syst. Zool.* 5:48–64.

Bruce-Chwatt, L. J. 1981. Malaria debated. *Nature* 294:302, 388.

Bulmer, M. 1980. *The Mathematical Theory of Quantitative Genetics*. Oxford: Clarendon Press.

Bull, J. J. 1983. *Evolution of Sex Determining Mechanisms*. Menlo Park, Calif.: Benjamin-Cummings Publ. Co.

Bull, J. J., and E. L. Charnov. 1985. On irreversible evolution. *Evolution* 39:1149–1155.

Bumpus, H. 1899. The elimination of the unfit as illustrated by the introduced sparrow, *Passer domesticus*. *Mar. Biol. Lab., Biol. Lect.* (Woods Hole, 1898), pp. 209–228.

Byron, E. R. 1982. The adaptive significance of calanoid copepod pigmentation: A comparative and experimental analysis. *Ecology* 63: 1871–1886.

Cain, A. J. 1964. The perfection of animals. Pp. 36–63 in J. D. Carthy and C. L. Duddington, eds., *Viewpoints in Biology*, vol. 3. London: Butterworths.

Cain, A. J. 1971. Colour and banding morphs in subfossil samples on the snail *Cepaea*. Pp. 65–92 in E. R. Creed, ed., *Ecological Genetics and Evolution, Essays in Honour of E. B. Ford*. Oxford: Blackwell Scientific Publications.

Cain, A. J. 1983. Ecology and ecogenetics of terrestrial molluscan populations. Pp. 597–647 in W. D. Russell-Hunter, ed., *The Mollusca*, vol. 6, *Ecology*. New York: Academic Press.

Cain, A. J., and J. D. Currey. 1968. Studies on *Cepaea*. III. Ecogenetics of a population of *Cepaea nemoralis* (L.) subject to strong area effects. *Phil. Trans. Roy. Soc. Lond. B.* 253:447–482.

Cain, A. J., and P. M. Sheppard. 1954. Natural selection in *Cepaea*. *Genetics* 39:89–116.

Calhoun, J. B. 1947. The role of temperature and natural selection in relation to the variations of the size of the English Sparrow in the United States. *Amer. Nat.* 81:203–228.

Cameron, R.A.D., K. Down, and D. J. Pannett. 1980. Historical and environmental influences on hedgerow snail faunas. *Biol. J. Linn. Soc. Lond.* 13:75–87.

Cameron, R.A.D., M. A. Carter, and M. A. Palles-Clark. 1980. *Cepaea* on Salisbury Plain: Patterns of variation, landscape history, and habitat stability. *Biol. J. Linn. Soc. Lond.* 14:335–358.

Camin, J. H., and P. R. Ehrlich. 1958. Natural selection in water snakes (*Natrix sipedon* L.) on islands in Lake Erie. *Evolution* 12:504–511.

Caplan, A. L. 1977. Tautology, circularity, and biological theory. *Amer. Nat.* 111:390–393.

Carter, M. A. 1967. Selection in mixed colonies of *Cepaea nemoralis* and *C. hortensis*. *Heredity* 22:117–139.

Carter, M. 1968. Thrush predation of an experimental population of the snail *Cepaea nemoralis* (L.). *Proc. Linn. Soc. Lond.* 179:241–249.

Castrodeza, C. 1977. Tautologies, beliefs, and empirical knowledge in biology. *Amer. Nat.* 111:393–394.

Cavalli-Sforza, L. L., and W. F. Bodmer. 1971. *The Genetics of Human Populations*. San Francisco: W. H. Freeman & Co.

Cavalli-Sforza, L. L., and M. W. Feldman. 1981. *Cultural Transmission and Evolution: A Quantitative Approach*. Princeton, N.J.: Princeton University Press.

Chakraborty, R., P. A. Fuerst, and M. Nei. 1980. Statistical studies on protein polymorphism in natural populations. III. Distribution of allele frequencies and the number of alleles per locus. *Genetics* 94:1039–1063.

Chapin, G., and R. Wasserstrom. 1981. Agricultural production and malaria resurgence in Central America and India. *Nature* 293:181–185.

Charlesworth, B. 1971. Selection in density-regulated populations. *Ecology* 52:469–474.

Charlesworth, B. 1980. *Evolution in Age-Structured Populations*. Cambridge, Eng.: Cambridge University Press.

Charlesworth, B., R. Lande, and M. Slatkin. 1982. A neo-Darwinian commentary on macroevolution. *Evolution* 36:474–498.

Charnov, E. L. 1982. *The Theory of Sex Allocation*. Princeton, N.J.: Princeton University Press.

Charnov, E. L., and J. R. Krebs. 1974. On clutch size and fitness. *Ibis* 116:217–219.

Charnov, E. L., and S. W. Skinner. 1984. Evolution of host selection and clutch size in parasitoid wasps. *Florida Entomologist* 67:5–21.

Choo, T. M., H. R. Klinck, and C. A. St.-Pierre. 1980a. The effect of location on natural selection in bulk populations of barley (*Hordeum vulgare* L.). I. Simply inherited traits. *Can. J. Plant. Sci.* 60:31–40.

Choo, T. M., H. R. Klinck, and C. A. St.-Pierre. 1980b. The effect of location on natural selection in bulk populations of barley (*Hordeum vulgare* L.). II. Quantitative traits. *Can. J. Plant. Sci.* 60:41–47.

Christiansen, F. B. 1977. Population genetics of *Zoarces viviparus* (L.): A review. Pp. 21–47 in F. B. Christiansen and O. Frydenberg, eds., *Measuring Selection in Natural Populations*. New York: Springer-Verlag.

Christiansen, F. B. 1980. Studies on selection components in natural populations using population samples of mother-offspring combinations. *Hereditas* 92:199–203.

Christiansen, F. B., J. Bundgaard, and J.S.F. Barker. 1977. On the structure of fitness estimates under post-observational selection. *Evolution* 31:843–853.

Christiansen, F. B., and O. Frydenberg. 1973. Selection component analysis of natural populations using population samples including mother-offspring combinations. *Theor. Pop. Biol.* 4:425–445.

Christiansen, F. B., and O. Frydenberg. 1976. Selection component analysis of natural populations using mother-offspring samples of successive cohorts. Pp. 277–301 in S. Karlin and E. Nevo, eds., *Population Genetics and Ecology*. New York: Academic Press.

Christiansen, F. B., O. Frydenberg, A. O. Gyldenholm, and V. Simonsen. 1974. Genetics of *Zoarces* populations. VI. Further evidence, based on age group samples, of a heterozygote deficit in the Est III polymorphism. *Hereditas* 77:225–235.

Christiansen, F. B., O. Frydenberg, and V. Simonsen. 1973. Genetics of *Zoarces* populations. IV. Selection component analysis of an esterase polymorphism using population samples including mother-offspring combinations. *Hereditas* 73:291–304.

Christiansen, F. B., O. Frydenberg, and V. Simonsen. 1977. Genetics of *Zoarces* populations. X. Selection component analysis of the Est III polymorphism using samples of successive cohorts. *Hereditas* 87:129–150.

Clark, A. G., and M. W. Feldman. 1981a. The estimation of epistasis in components of fitness in experimental populations of *Drosophila melanogaster*. II. Assessment of meiotic drive, viability, fecundity, and sexual selection. *Heredity* 46:347–377.

Clark, A. G., and M. W. Feldman. 1981b. Disequilibrium between linked inversions: An alternative hypothesis. *Heredity* 46:379–390.

Clark, A. G., M. W. Feldman, and F. B. Christiansen. 1981. The estimation of epistasis in components of fitness in experimental populations of *Drosophila melanogaster*. I. A two-stage maximum likelihood model. *Heredity* 46:321–346.

Clark, P. J., and J. N. Spuhler. 1959. Differential fertility in relation to body dimensions. *Human Biol.* 31:121–137.

Clarke, B. C. 1972. Density-dependent selection. *Amer. Nat.* 106:1–13.

Clarke, B. C. 1975. The contribution of ecological genetics to evolutionary theory: Detecting the direct effects of natural selection on particular polymorphic loci. *Genetics* (supplement) 79:101–113.

Clarke, B. C. 1978. Some contributions of snails to the development of ecological genetics. Pp. 159–170 in P. F. Brussard, ed., *Ecological Genetics: The Interface*. New York: Springer-Verlag.

Clarke, B. C., W. Arthur, D. T. Horsley, and D. T. Parkin. 1978. Genetic variation and natural selection in pulmonate molluscs. Pp. 219–270 in V. Fetter and J. Peake, eds., *The Pulmonates*, vol. 2a, *Systematics, Evolution and Ecology*. New York: Academic Press.

Clarke, B. C., and J. J. Murray. 1962. Changes of gene frequency in *Cepaea nemoralis*. II. The estimation of selective values. *Heredity* 17:467–476.

Clarke, C. A., and P. M. Sheppard. 1966. A local survey of the distribution of industrial melanic forms in the moth *Biston betularia* and estimates of the selective values of those in an industrial environment. *Proc. Roy. Soc. Lond. B.* 165:424–439.

Clegg, M. T., and R. W. Allard. 1972. Patterns of genetic differentiation in the slender wild oat species *Avena barbata*. *Proc. Nat. Acad. Sci. U.S.A.* 69:1820–1824.

Clegg, M. T., and R. W. Allard 1973. Viability versus fecundity selection in the slender wild oat *Avena barbata* L. *Science* 181:667–668.

Clegg, M. T., A. L. Kahler, and R. W. Allard. 1978a. Estimation of life cycle components of selection in an experimental plant population. *Genetics* 89:765–792.

Clegg, M. T., A. L., Kahler, and R. W. Allard. 1978b. Genetic demography of plant populations. Pp. 173–188 in P. F. Brussard, ed., *Ecological Genetics: The Interface*. New York: Springer-Verlag.

Clutton-Brock, T. H. 1983. Selection in relation to sex. Pp. 457–481 in D. S. Bendall, ed., *Evolution from Molecules to Men*. Cambridge, Eng.: Cambridge University Press.

Clutton-Brock, T. H., F. E. Guinness, and S. D. Albon. 1982. *Red Deer: Behavior and Ecology of Two Sexes*. Chicago: University of Chicago Press.

Cody, M. L., and J. H. Brown. 1970. Character convergence in Mexican finches. *Evolution* 24:304–310.

Cody, M. L., and H. A. Mooney. 1978. Convergence versus nonconvergence in Mediterranean-climate ecosystems. *Ann. Rev. Ecol. Syst.* 9:265–321.

Colwell, R. K., and D. W. Winkler. 1984. A null model for null models in biogeography. Pp. 344–359 in D. R. Strong, D. Simberloff, L. G. Abele, and A. B. Thistle, eds., *Ecological Communities: Conceptual Issues and the Evidence*. Princeton, N.J.: Princeton University Press.

Compton, S. G., S. G. Beesley, and D. A. Jones. 1983. On the polymorphism of cyanogenesis in *Lotus corniculatus* L. IX. Selective herbivory in natural populations at Porthdafarch, Anglesey. *Heredity* 51:537–547.

Conroy, B. A., and J. A. Bishop. 1980. Maintenance of the polymorphism for melanism in the moth *Phigalia pilosaria* in rural North Wales. *Proc. Roy. Soc. Lond. B.* 210:285–298.

Cook, A. D., P. R. Atsatt, and C. A. Simon. 1971. Doves and dove weed: Multiple defenses against avian predators. *BioScience* 21:277–281.

Cook, L. M. 1971. *Coefficients of Natural Selection.* London: Hutchinson University Library.

Cook, L. M., and G. S. Mani. 1980. A migration-selection model for the morph frequency variation in the peppered moth over England and Wales. *Biol. J. Linn. Soc. Lond.* 13:179–198.

Cooke, F., and C. S. Findlay. 1982. Polygenic variation and stabilizing selection in a wild population of lesser snow geese (*Anser caerulescens*). *Amer. Nat.* 120:543–550.

Cooper, W. S. 1984. Expected time to extinction and the concept of fundamental fitness. *J. Theor. Biol.* 107:603–629.

Cooper, W. S., and R. H. Kaplan. 1982. Adaptive "coin-flipping": A decision-theoretic examination of natural selection for random individual variation. *J. Theor. Biol.* 94:135–151.

Cooper-Driver, G. A., and T. Swain. 1976. Cyanogenic polymorphism in bracken in relation to herbivore predation. *Nature* 260:604.

Cornfield, J. 1962. Joint dependence of risk of coronary heart disease on serum cholesterol and systolic blood pressure: A discriminant function analysis. *Federation Proc.* 21:58–61.

Cox, B. 1981. Origin of species: Premises, premises. *Nature* 291:373.

Cox, D. R. 1972. Regression models and life tables. *J. Roy. Statistical Soc. B.* 34:187–202, discussion pp. 202–220.

Crampton, H. E. 1904. Experimental and statistical studies upon lepidoptera. I. Variation and elimination in *Philosamia cynthia. Biometrika* 3:113–130.

Crandall, R. E., S. C. Stearns, and J. R. Dudman. 1985. An evaluation of fitness measures for whole organisms: A combinatorial approach. Manuscript.

Crawford-Sidebotham, T. J. 1972. The role of slugs and snails in the maintenance of the cyanogenic polymorphisms of *Lotus corniculatus* and *Trifolium repens. Heredity* 28:405–411.

Creed, E. R. 1971. Industrial melanism in the two-spot ladybird and smoke abatement. *Evolution* 25:290–293.

Creed, E. R., D. R. Lees, and M. G. Bulmer. 1980. Pre-adult viability differences of melanic *Biston betularia* (L.) (Lepidoptera). *Biol. J. Linn. Soc. Lond.* 13:251–262.

Crossner, K. A. 1977. Natural selection and clutch size in the European starling. *Ecology* 58:885–892.

Crow, J. F. 1958. Some possibilities for measuring selection intensities in man. *Human Biol.* 30:1–13.

Crow, J. F. 1979. Minor viability mutants in *Drosophila*. *Genetics* (supplement) 92:165–172.

Crow, J. F., and M. Kimura. 1979. Efficiency of truncation selection. *Proc. Nat. Acad. Sci. U.S.A.* 76:396–399.

Curtis, C. F. 1981. Malaria debated. *Nature* 294:388.

Daday, H. 1954a. Gene frequencies in strains of *Trifolium repens* L. *Nature* 174:521.

Daday, H. 1954b. Gene frequencies in wild populations of *Trifolium repens*. I. Distribution by latitude. *Heredity* 8:61–78.

Daday, H. 1958. Gene frequencies in wild populations of *Trifolium repens*. III. World distribution. *Heredity* 12:169–184.

Daday, H. 1965. Gene frequencies in wild populations of *Trifolium repens*. IV. Mechanisms of natural selection. *Heredity* 20:355–365.

Darlington, P. J., Jr. 1983. Evolution: Questions for the modern theory. *Proc. Nat. Acad. Sci. U.S.A.* 80:1960–1963.

Darwin, C. 1859. *On The Origin of Species by Means of Natural Selection, or the Preservation of Favoured Races in the Struggle for Life*. London: John Murray.

Darwin, C. 1869. Fifth edition of Darwin 1859.

Darwin, C. 1871. *The Descent of Man, and Selection in Relation to Sex*. 2 vols. London: John Murray.

Darwin, C. 1875. *The Variation of Animals and Plants Under Domestication*. 2d rev. ed. London: John Murray.

David, J. R., J. Van Herrewege, M. Monclus, and A. Prevosti. 1979. High ethanol tolerance of two distantly related *Drosophila* species: A probable case of recent convergent adaptation. *Comp. Biochem. Physiol.* 63C:53–56.

Davidson, D. W. 1982. Sexual selection in harvester ants (Hymenoptera: Formicidae: *Pogonomyrmex*). *Beh. Ecol. Sociobiol.* 10:245–250.

Davis, G. M. 1979. The origin and evolution of the gastropod family Pomatiopsidae, with emphasis on the Mekong river Triculinae. Monograph 20, Academy of Natural Sciences of Philadelphia.

Dawkins, R. 1976. *The Selfish Gene*. Oxford: Oxford University Press.

Dawkins, R. 1982. *The Extended Phenotype: The Gene as the Unit of Selection*. San Francisco: W. H. Freeman & Co.

Day, P. R. 1974, *Genetics of Host-Parasite Interactions*. San Francisco: W. H. Freeman.

de Jong, G. 1982. Fecundity selection and maximization of equilibrium number. *Neth. J. Zool.* 32:572–595.

Dhondt, A. A., R. Eyckerman, and J. Huble. 1979. Will great tits become little tits? *Biol. J. Linn. Soc. Lond.* 11:289–294.

DiCesnola, A. P. 1904. Preliminary notes on the protective value of colour in *Mantis religiosa*. *Biometrika* 3:58–59.

DiCesnola, A. P. 1907. A first study of natural selection in *Helix arbustorum* (Helicogena). *Biometrika* 5:387–399.

Dietz, E. J. 1983. Permutation tests for association between two distance matrices. *Sys. Zool.* 32:21–26.

DiMichele, L., and D. A. Powers. 1982a. LDH-B genotype-specific hatching times of *Fundulus heteroclitus* embryos. *Nature* 296:563–564.

DiMichele, L., and D. A. Powers. 1982b. Physiological basis for swimming endurance differences between LDH-B genotypes of *Fundulus heteroclitus*. *Science* 216:1014–1016.

DiMichele, L., and D. A. Powers. 1984. Developmental and oxygen consumption rate differences between lactate dehydrogenase-B genotypes of *Fundulus heteroclitus* and their effect on hatching time. *Physiol. Zool.* 57:52–56.

Dirzo, R., and J. L. Harper. 1982a. Experimental studies on slug-plant interactions. III. Differences in the acceptability of individual plants of *Trifolium repens* to slugs and snails. *J. Ecol.* 70:101–117.

Dirzo, R., and J. L. Harper. 1982b. Experimental studies on slug-plant interactions. IV. The performance of cyanogenic and acyanogenic morphs of *Trifolium repens* in the field. *J. Ecol.* 70:119–138.

Dobzhansky, Th. 1941. *Genetics and the Origin of Species*. New York: Columbia University Press.

Dobzhansky, Th. 1956. What is an adaptive trait? *Amer. Nat.* 90: 337–347.

Dobzhansky, Th. 1968a. On some fundamental concepts of Darwinian biology. *Evol. Biol.* 2:1–34.

Dobzhansky, Th. 1968b. Adaptedness and fitness. Pp. 109–121 in

R. C. Lewontin, ed., *Population Biology and Evolution*. Syracuse, N.Y.: Syracuse University Press.

Dobzhansky, Th. 1970. *Genetics of the Evolutionary Process*. New York: Columbia University Press.

Dobzhansky, Th., and O. Pavlovsky. 1953. Indeterminate outcome of certain experiments on *Drosophila* populations. *Evolution* 7:198–210.

Dolinger, P. M., P. R. Ehrlich, W. L. Fitch, and D. E. Breedlove. 1973. Alkaloid and predation patterns in Colorado lupine populations. *Oecologia Berl.* 13:191–204.

Douglas, M. E., and J. A. Endler. 1983. Quantitative matrix comparisons in ecological and evolutionary investigations. *J. Theor. Biol.* 99:777–795.

Douglas, M. M., and J. W. Grula. 1978. Thermoregulatory adaptations allowing ecological range expansion by the pierid butterfly *Nathalis iole* Boisduval. *Evolution* 32:776–783.

Dowdeswell, W. H. 1961. Experimental studies on natural selection in the butterfly *Maniola jurtina*. *Heredity* 16:39–52.

Dowdeswell, W. H. 1962. A further study of the butterfly *Maniola jurtina* in relation to natural selection by *Apanteles tetricus*. *Heredity* 17:513–523.

Downhower, J. F., and L. Brown. 1980. Mate preferences of female mottled sculpins, *Cottus bairdi*. *Anim. Behav.* 28:728–734.

Dritschilo, W., and D. Pimintel. 1979. Unpublished results quoted on p. 55 of W. Dritschilo, J. Krummel, D. Nafus, and D. Pimentel, Herbivorous insects colonising cyanogenic and acyanogenic *Trifolium repens*. *Heredity* 42:49–56.

Dunbar, R.I.M. 1982. Adaptation, fitness, and the evolutionary tautology. Pp. 9–28 in King's College Sociolbiology Group, eds., *Current Problems in Sociobiology*. Cambridge, Eng.: Cambridge University Press.

Dunham, A. E., G. R. Smith, and J. N. Taylor. 1979. Evidence for ecological character displacement in western American Catostomid fishes. *Evolution* 33:877–896.

Eanes, W. F., P. M. Gaffney, R. K. Koehn, and C. M. Simon. 1977. A study of sexual selection in natural populations of the milkweed beetle, *Tetraopes tetraophthalmus*. Pp. 49–64 in F. B. Christiansen and O. Frydenberg, eds. *Measuring Selection in Natural Populations*. New York: Springer-Verlag.

REFERENCES

Edmunds, G. F., Jr., and D. N. Alstad. 1978. Coevolution in insect herbivores and conifers. *Science* 199:941–945.

Edmunds, G. F., Jr., and D. N. Alstad. 1981. Responses of black pineleaf scales to host plant variability. Pp. 29–38 in R. F. Denno and H. Dingle, eds., *Insect Life History Patterns, Habitat and Geographic Variation*. New York: Springer-Verlag.

Edmunds, M. 1966. Natural selection in the mimetic butterfly *Hypolimnas misippus* L. in Ghana. *Nature* 212:1478.

Edmunds, M. 1974. Significance of beak marks on butterfly wings. *Oikos* 25:117–118.

Ehrlich, P. R., and J. H. Camin. 1960. Natural selection on middle island water snakes (*Natrix sipedon* L.). *Evolution* 14:136.

Ehrlich, P. R., and L. G. Mason. 1966. The population biology of the butterfly *Euphydryas editha*. III. Selection and the phenetics of the Jasper Ridge colony. *Evolution* 20:165–173.

Eldredge, N., and J. Cracraft. 1980. *Phylogenetic Patterns and the Evolutionary Process*. New York: Columbia University Press.

Ellis, W. M., R. J. Keymer, and D. A. Jones. 1977a. The defensive function of cyanogenesis in natural populations. *Experientia* 33:309–310.

Ellis, W. M., R. J. Keymer, and D. A. Jones. 1977b. On the polymorphism of cyanogenesis in *Lotus corniculatus* L. VIII. Ecological studies in Anglesy. *Heredity* 39:45–65.

Ellis, W. M., R. J. Keymer, and D. A. Jones. 1977c. The effect of temperature on the polymorphism of cyanogenesis in *Lotus corniculatus* L. *Heredity* 38:339–347.

Elston, R. C., and R. Forthofer. 1977. Testing for Hardy-Weinberg equilibrium in small samples. *Biometrics* 33:536–542.

Emberton, L.R.B., and S. Bradbury. 1963. Transmission of light through shells of *Cepaea nemoralis* L. *Proc. Malacol. Soc. Lond.* 35:211–219.

Emerson, S. B., and L. Radinsky. 1980. Functional analysis of sabertooth cranial morphology. *Paleobiology* 6:295–312.

Emmons, L. H., and A. H. Gentry. 1983. Tropical forest structure and the distribution of gliding and prehensile-tailed vertebrates. *Amer. Nat.* 121:513–524.

Endler, J. A. 1973. Gene flow and population differentiation. *Science* 179:243–250.

Endler, J. A. 1977. *Geographic Variation, Speciation, and Clines*. Princeton, N.J.: Princeton University Press.

Endler, J. A. 1978. A predator's view of animal color patterns. *Evol. Biol.* 11:319–364.

Endler, J. A. 1980. Natural selection on color patterns in *Poecilia reticulata*. *Evolution* 34:76–91.

Endler, J. A. 1982a. Problems in distinguishing historical from ecological factors in biogeography. *Amer. Zool.* 22:441–452.

Endler, J. A. 1982b. Convergent and divergent effects of natural selection on color patterns in two fish faunas. *Evolution* 36:178–188.

Endler, J. A. 1983. Natural selection on color patterns in poeciliid fishes. *Envir. Biol. Fishes* 9:173–190.

Endler, J. A. 1984. Progressive background matching in moths and a quantitative measure of crypsis. *Biol. J. Linn. Soc. Lond.* 22:187–231.

Ennos, R. A. 1981. Detection of selection in populations of white clover (*Trifolium repens* L.). *Biol. J. Linn. Soc. Lond.* 15:75–82.

Etges, W. J. 1984. Genetic structure and change in natural populations of *Drosophila robusta*: Systematic inversion and inversion associated frequency shifts in the Great Smokey Mountains. *Evolution* 38:675–688.

Ewens, W. J. 1972. The sampling theory of selectively neutral alleles. *Theor. Popul. Biol.* 3:87–112.

Ewens, W. J. 1979. *Mathematical Population Genetics*. New York: Springer-Verlag.

Ewens, W., and M. W. Feldman. 1976. The theoretical assessment of selective neutrality. Pp. 303–337 in S. Karlin and E. Nevo, eds., *Population Genetics and Ecology*. New York: Academic Press.

Falconer, D. S. 1981. *Introduction to Quantitative Genetics*. 2d ed. London: Longmans.

Farentinos, R. C., P. J. Capretta, R. E. Kepner, and V. M. Littlefield. 1981. Selective herbivory in tassel-eared squirrels: Role of monoterpenes in ponderosa pines chosen as feeding trees. *Science* 213:1273–1275.

Farewell, V. T., and S. Dahlberg. 1983. On the comparison of procedures for testing the equality of survival curves. *Biometrika* 70:707–709.

Fawcett, M. H. 1984. Local and latitudinal variation in predation on an herbivorous marine snail. *Ecology* 65:1214–1230.

Feaster, C. V., E. F. Young, Jr., and E. L. Turcotte. 1980. Comparison of artificial and natural selection in American Pima cotton under different environments. *Crop. Sci.* 20:555–558.

Feldman, M. W., and F. B. Christiansen. 1975. The effect of population subdivision on two loci without selection. *Genet. Res. Cambr.* 24:151–162.

Feller, W. 1968. *An Introduction to Probability Theory and Its Applications.* 3d ed., vol. 1. New York: Wiley.

Fenner, F., and F. N. Ratcliffe. 1965. *Myxomatosis.* Cambridge, Eng.: Cambridge University Press.

Ferguson, G. W., and S. F. Fox. 1984. Annual variation of survival advantage of large juvenile side-blotched lizards, *Uta stansburiana*: Its causes and evolutionary significance. *Evolution* 38:342–349.

Finch, V. A., and D. Western. 1977. Cattle colors in pastoral herds: Natural selection or social preference? *Ecology* 58:1384–1392.

Fisher, R. A. 1930. *The Genetical Theory of Natural Selection.* Oxford: Oxford University Press.

Fisher, R. A. 1939. Selective forces in wild populations of *Paratettix texanus. Ann. Eugen.* 9:109–122.

Fleischer, R. C., and R. F. Johnston. 1982. Natural selection on body size and proportions in house sparrows. *Nature* 298:747.

Fleischer, R. C., and R. F. Johnston. 1984. The relationships between winter climate and selection on body size of house sparrows. *Canad. J. Zool.* 62:405–410.

Flew, A. 1981. Fight the obscure. *Nature* 292:192.

Flor, H. H. 1956. The complementary genic systems in flax and flax rust. *Adv. Genet.* 8:29–54.

Fong, D. W. 1985. A quantitative genetic analysis of regressive evolution in the amphipod *Gammarus minus* Say. Ph.D. dissertation, Northwestern Universtiy, Evanston, Illinois.

Fontdevila, A., C. Zapata, G. Alvarez, L. Sanchez, J. Mendez, and I. Enriquez. 1983. Genetic coadaptation in the chromosomal polymorphism of *Drosophila subobscura.* I. Seasonal changes of gametic disequilibrium in a natural population. *Genetics* 105:935–955.

Ford, E. B. 1975. *Ecological Genetics.* 4th ed. London: Chapman & Hall.

Ford, E. B., and P. M. Sheppard. 1969. The *medionigra* polymorphism of *Panaxia dominula. Heredity* 24:561–659.

Foulds, W. 1977. The physiological response to moisture supply of cyanogenic and acyanogenic phenotypes of *Lotus corniculatus* L. and *Trifolium repens* L. *Heredity* 39:219–234.

Foulds, W., and J. P. Grime. 1972. The response of cyanogenic and acyanogenic phenotypes of *Trifolium repens* to soil moisture supply. *Heredity* 28:181–187.

Foulds, W., and L. Young. 1977. Effects of frosting, moisture stress, and potassium cyanide on the metabolism of cyanogenic and acyanogenic phenotypes of *Lotus corniculatus* L. and *Trifolium repens* L. *Heredity* 38:19–24.

Fox, S. F. 1975. Natural selection on morphological phenotypes of the lizard *Uta stansburiana*. *Evolution* 29:95–107.

Fuerst, P. A., R. Chakraborty, and M. Nei. 1977. Statistical studies on protein polymorphism in natural populations. I. Distribution of single locus heterozygosity. *Genetics* 86:455–483.

Fujino, K., and T. Kang. 1968. Transferrin groups of tuna. *Genetics* 59:79–91.

Gallagher, J. C. 1980. Population genetics of *Skeletonema costatum* (Bacillarcophyceae) in Narragansett Bay. *J. Phycol.* 16:464–474.

Gallun, R. L., K. J. Starks, and W. D. Guthrie. 1975. Plant resistance to insects attacking cereals. *Ann. Rev. Entomol.* 20:337–357.

Gans, C. 1974. *Biomechanics, an Approach to Vertebrate Biology*. Philadelphia: J. B. Lippincott Co.

Gardner, W. D., and L. G. Sullivan. 1981. Benthic storms: Temporal variability in a deep-ocean nepheloid layer. *Science* 213:329–331.

Gartside, D. W., and T. McNeilly. 1974. The potential for evolution of heavy metal tolerance in plants. *Heredity* 32:335–348.

Gendron, R. P. 1981. Survivalist hopes. *Nature* 293:602.

Georghiou, G. P. 1972. The evolution of resistance to pesticides. *Ann. Rev. Ecol. Syst.* 3:133–168.

Gerould, J. H. 1921. Blue-green caterpillars: The origin and ecology of a mutation in hemolymph color in *Colias* (*Eurymus*) *philodice*. *J. Exp. Zool.* 34:385–415.

Gerould, J. H. 1926. Inheritance of olive-green and blue-green variations appearing in the life-cycle of a butterfly *Colias philodice*. *J. Exp. Zool.* 43:413–427.

Ghiselin, M. T. 1969. *The Triumph of the Darwinian Method*. Berkeley: University of California Press.

Ghiselin, M. T. 1974. *The Economy of Nature and the Evolution of Sex.* Berkeley: University of California Press.

Ghiselin, M. T. 1981. Categories, life, and thinking. *Behavioral and Brain Sciences* 4:269–313.

Giesel, J. T. 1970. On the maintenance of shell pattern and behavior polymorphism in *Acmaea digitalis*, a limpet. *Evolution* 24:98–119.

Gill, P. 1981. Allozyme variation in sympatric populations of British grasshoppers—evidence of natural selection. *Biol. J. Linn. Soc.* (London) 16:83–91.

Gillespie, J. H. 1973. Polymorphism in random environments. *Theor. Popul. Biol.* 4:193–195.

Gillespie, J. H. 1977. Natural selection for offspring numbers: A new evolutionary principle. *Amer. Nat.* 111:1010–1014.

Gingerich, P. D. 1983. Rates of evolution: Effect of time and temporal scaling. *Science* 222:159–161.

Givnish, T. 1979. On the adaptive significance of leaf form. Pp. 375–407 in O. T. Solbrig, S. Jain, G. B. Johnson, and P. H. Raven, eds., *Topics in Plant Population Biology.* New York: Columbia University Press.

Goldstein, M., and W. R. Dillon. 1978. *Discrete Discriminant Analysis.* New York: Wiley.

Goodman, M. M. 1974. Genetic distances: Measuring dissimilarity among populations. *Yearbook of Physical Anthropology* 17:1–38.

Gordon, C. 1935. An experiment on a released population of *Drosophila melanogaster. Amer. Nat.* 69:381.

Gould, F. 1979. Rapid host range evolution in a population of the phytophagous mite *Tetranychus urtricae* Koch. *Evolution* 33:791–802.

Gould, S. J. 1980. Is there a new and general theory of evolution emerging? *Paleobiology* 6:119–130.

Gould, S. J. 1982. Darwinism and the expansion of evolutionary theory. *Science* 216:380–387.

Gould, S. J. 1984. Covariance sets and ordered geographic variation in *Cerion* from Aruba, Bonaire and Curaçao: A way of studying nonadaptation. *Syst. Zool.* 33:217–237.

Gould, S. J., and R. C. Lewontin. 1979. The spandrels of San Marco and the panglossian paradigm: A critique of the adaptationist programme. *Proc. Roy. Soc. Lond. B.* 205:581–598.

Gould, S. J., and E. S. Vrba. 1982. Exaptation—a missing term in the science of form. *Paleobiology* 8:14–15.

Goux, J. M., and D. Anxolabehere. 1980. The measurement of sexual isolation and selection: A critique. *Heredity* 45:255–262.

Gowen, J. W. 1963. Pp. 163–194 in W. J. Schull, ed., *Genetic Selection in Man*. Ann Arbor: University of Michigan Press.

Grant. B. R. 1985. Selection on bill characters in a population of Darwin's finches: *Geospiza conirostris* on Isla Genovesa, Galapagos. *Evolution* 39:523–532.

Grant, P. R. 1972a. Convergent and divergent character displacement. *Biol. J. Linn. Soc. Lond.* 4:39–68.

Grant, P. R. 1972b. Centripetal selection and the house sparrow. *Syst. Zool.* 21:23–30.

Grant, P. R. 1975. The classical case of character displacement. *Evol. Biol.* 8:237–337.

Grant, P. R., B. R. Grant, J.N.M. Smith, I. J. Abbott, and L. K. Abbott. 1976. Darwin's finches: Population variation and natural selection. *Proc. Nat. Acad. Sci. U.S.A.* 13:257–261.

Grant, P. R., and T. D. Price. 1981. Population variation in continuously varying traits as an ecological genetics problem. *Amer. Zool.* 21:795–811.

Greaves, J. H., and P. Ayers. 1967. Heritable resistance to Warfarin in rats. *Nature* 215:877–878.

Greaves, J. H., and P. Ayers. 1969. Linkage between genes for coat color and resistance to Warfarin in *Rattus norvegicus*. *Nature* 224: 284–285.

Gross, H. P. 1978. Natural selection by predators on the defensive apparatus of the three-spined stickleback, *Gasterosteus aculeatus* L. *Canad. J. Zool.* 56:398–413.

Hagen, D. W. 1973. Inheritance of numbers of lateral plates and gill rakers in *Gasterosteus aculeatus*. *Heredity* 30:303–312.

Hagen, D. W., and L. G. Gilbertson. 1972. Geographic variation and environmental selection in *Gasterosteus aculeatus* L. in the Pacific Northwest, America. *Evolution* 26:32–51.

Hagen, D. W., and L. G. Gilbertson. 1973a. The genetics of plate morphs in freshwater three-spine sticklebacks. *Heredity* 31:75–84.

Hagen, D. W., and L. G. Gilbertson. 1973b. Selective predation and the intensity of selection acting upon the lateral plates of three-spine sticklebacks. *Heredity* 30:273–287.

Hagen, D. W., and J. D. McPhail. 1970. The species problem within *Gasterosteus aculeatus* on the Pacific coast of North America. *J. Fish. Res. Bd. Canada* 27:147–155.

Hagen, D. W., G.E.E. Moodie, and P. F. Moodie. 1980. Polymorphism for breeding colors in *Gasterosteus aculeatus*. II. Reproductive success as a result of convergence for threat display. *Evolution* 34: 1050–1059.

Hailman, J. P. 1982. Evolution and behavior: An iconoclastic view. Pp. 205–254 in H. C. Plotkin, ed., *Learning, Development, and Culture*. New York: Wiley.

Hakkinen, H., and S. Koponen. 1982. The marginal populations of *Cepaea hortensis*: Morph frequencies, shell size, and predation. *Hereditas* 97:163–166.

Haldane, J.B.S. 1935. Darwinism under revision. *Rationalist Ann.* (1934), pp. 19–29.

Haldane, J.B.S. 1954. The measurement of natural selection. *Proc. IX Intl. Congr. Genetics* (*Caryologica* supplement) 1:480–487.

Haldane, J.B.S., and S. D. Jayakar. 1963. Polymorphism due to selection in varying directions. *J. Genetics*. 58:237–242.

Halkka, O., and M. Raatikainen. 1975. Transfer of individuals as a means of investigating natural selection in operation. *Hereditas* 80: 27–34.

Halkka, O., M. Raatikainen, L. Halkka, and R. Hovinen. 1975. The genetic composition of *Philaenus spumarius* populations in island habitats variably affected by voles. *Evolution* 29:700–706.

Hamrick, J. L., and R. W. Allard. 1975. Correlations between quantitative characters and enzyme genotypes in *Avena barbata*. *Evolution* 29:438–442.

Hamrick, J. L., and L. R. Holden. 1979. Influence of microhabitat heterogeneity on gene frequency distribution and gametic phase disequilibrium in *Avena barbata*. *Evolution* 33:521–533.

Harding, J. 1970. Genetics of *Lupinus*. II. The selective disadvantage of the pink flower color mutant in *Lupinus nanus*. *Evolution* 24:120–127.

Harland, S. C. 1947. An alteration in gene frequency in *Ricinus communis* L. due to climatic conditions. *Heredity* 1:121–125.

Harrison, R. G. 1977. Parallel variation at an enzyme locus in sibling species of field crickets. *Nature* 266:168–170.

Hartley, B. S. 1979. Evolution of enzyme structure. *Proc. Roy. Soc. Lond. B.* 205:443–452.

Harvey, P. H., N. Birley, and T. H. Blackstock. 1975. The effect of experience on the selective behavior of song thrushes feeding on artificial populations of *Cepaea* (Held). *Genetica* 45:211–216.

Harvey, P. H., R. K. Colwell, J. W. Silverton, and R. M. May. 1983. Null models in ecology. *Ann. Rev. Ecol. Syst.* 14:189–211.

Haskins, C. P., E. F. Haskins, J.J.A. McLaughlin, and R. E. Hewitt. 1961. Polymorphism and population structure in *Lebistes reticulatus*, a population study. Pp. 320–395 in W. F. Blair, ed., *Vertebrate Speciation*. Austin: University of Texas Press.

Heath, D. J. 1974. Seasonal changes in frequency of the "yellow" morph of the isopod *Sphaeroma rugicauda. Heredity* 32:299–307.

Heath, D. J. 1975. Colour, sunlight, and internal temperatures in the land-snail *Cepaea nemoralis* (L.). *Oecologia Berl.* 19:29–38.

Hedrick, P. W. 1975. Genetic similarity and distance: Comments and comparisons. *Evolution* 29:362–366.

Hedrick, P. W. 1983. *Genetics of Populations*. New York: Van Nostrand Reinhold Co.

Hedrick, P. W., and L. Holden. 1979. Hitch-hiking: An alternative to coadaptation for the barley and slender wild oat examples. *Heredity* 43:79–86.

Hegmann, J. P., and H. Dingle. 1982. Phenotypic and genetic covariance structure in milkweed bug life history traits. Pp. 177–185 in H. Dingle and J. P. Hegmann, eds., *Evolution and Genetics of Life Histories*. New York: Springer-Verlag.

Hickey, D. A., and T. McNeilly. 1975. Competition between metal tolerant and normal plant populations: A field experiment on normal soil. *Evolution* 29:458–464.

Hilbish, T. J., and R. K. Koehn. 1985. The physiological basis of natural selection at the *Lap* locus. *Evolution* 39:1302–1317.

Hill, W. G. 1982a. Predictions of response to artificial selection from new mutations. *Genet. Res. Cambr.* 40:255–278.

Hill, W. G. 1982b. Rates of change in quantitative traits from fixation of new mutations. *Proc. Nat. Acad. Sci. U.S.A.* 79:142–145.

Hill, W. G., and A. Robertson. 1968. Linkage disequilibrium in finite populations. *Theor. Appl. Genet.* 38:226–231.

Hiorns, R. W., and G. A. Harrison. 1970. Sampling for the detection of natural selection by age group differences. *Human Biol.* 43:53–64.

Hoffman, R. J., and W. B. Watt. 1974. Naturally occurring variation in larval color of *Colias* butterflies: Isolation from two Colorado populations. *Evolution* 28:326–332.

Holliday, R. J., and P. D. Putwain. 1980. Evolution of herbicide resistance in *Senecio vulgaris*: Variation in susceptibility to simazine between and within populations. *J. Appl. Ecol.* 17:779–791.

Howard, R. D. 1979. Estimating reproductive success in natural populations. *Amer. Nat.* 114:221–231.

Howard, R. D. 1980. Mating behaviour and mating success in woodfrogs, *Rana sylvatica*. *Anim. Behav.* 28:705–716.

Howard, R. D., and A. G. Kluge. 1985. Proximate mechanisms of sexual selection in woodfrogs. *Evolution* 39:260–277.

Hull, D. L. 1980. Individuality and selection. *Ann. Rev. Ecol. Syst.* 11:311–332.

Hutt. F. B. 1963. Pp. 200–206 in W. J. Schull, ed., *Genetic Selection in Man*. Ann Arbor: University of Michigan Press.

Hutt, F. B., and R. D. Crawford. 1960. On breeding chicks resistant to pullorum disease without exposure thereto. *Can. J. Genet. Cytol.* 2:357–370.

Imai, T., and A. A. Gomez. 1979. Differentiation in sorghum varieties caused by tropical and temperate environments. *Japan Agricultural Research Quarterly* 13:149–151.

Inger, R. F. 1942. Differential selection of variant juvenile snakes. *Amer. Nat.* 76:104–109.

Inger, R. F. 1943. Further notes on differential selection of variant juvenile snakes. *Amer. Nat.* 77:87–90.

Ingram, V. M. 1957. Gene mutations in human haemoglobin: The chemical difference between normal and sickle-cell haemoglobin. *Nature* 180:326–328.

Jablonski, D., and R. A. Lutz. 1983. Larval ecology of marine benthic insects: Palaeobiological implications. *Biol. Rev.* 58:21–89.

Jain, S. K., and R. W. Allard. 1960. Population studies in predominantly self-pollinated species. I. Evidence for heterozygote advantage in a closed population of barley. *Proc. Nat. Acad. Sci. U.S.A.* 46:1371–1377.

Jain, S. K., and P. S. Martins. 1979. Ecological genetics of the colonizing ability of rose clover (*Trifolium hirtum* All.). *Amer. J. Bot.* 66:361–366.

Jain, S. K., and K. N. Rai. 1977. Natural selection during germination in wild oats (*Avena fatua*) and California burr clover (*Medicago polymorphia* var. *vulgaris*) populations. *Weed Sci.* 25:495–498.

Jain, S. K., and K. N. Rai. 1980. Population biology of *Avena*. VIII. Colonization experiment as a test of the role of natural selection in population divergence. *Amer. J. Bot.* 67:1342–1346.

Jaksić, F. M., and E. R. Fuentes. 1980. Correlation of tail loss in twelve species of *Liolaemus* lizards. *J. Herpetol.* 14:137–141.

Jaksić, F. M., and H. W. Greene. 1984. Empirical evidence of non-correlation between tail loss frequency and predation intensity on lizards. *Oikos* 42:407–411.

Jameson, D. L., and S. Pequegnat. 1971. Estimation of relative viability and fecundity of color polymorphisms in Anurans. *Evolution* 25:180–194.

Johnson, C. 1976. *Introduction to Natural Selection.* Baltimore, Md. University Park Press.

Johnson, D. M., G. L. Steward, M. Corley, R. Ghrist, J. Hagner, A. Ketterer, B. McDonell, W. Newsom, E. Owen, and P. Samuels. 1980. Brown-headed Cowbird (*Molothrus ater*) mortality in an urban winter roost. *Auk* 97:299–320.

Johnson, G. 1976. Polymorphism and predictability at the α-Gpdh locus in *Colias* butterflies. *Biochem. Genet.* 14:403–425.

Johnson, M. S. 1974. Comparative geographic variation in *Menidia*. *Evolution* 28:607–618.

Johnson, M. S. 1981. Effects of migration and habitat choice on shell banding frequencies in *Theba pisana* at a habitat boundary. *Heredity* 47:121–133.

Johnson, S. 1750, 1751 (reprinted 1903). *The Rambler.* From *The Complete Works of Samuel Johnson.* Troy, N.Y.: Pafraets Book Co.

Johnston, R. F. 1973. Evolution in the house sparrow. IV. Replicate studies in phenetic covariation. *Syst. Zool.* 22:219–226.

Johnston, R. F., and R. C. Fleischer. 1981. Overwinter mortality and sexual size dimorphism in the house sparrow. *Auk* 98:503–511.

Johnston, R. F., D. M. Niles, and S. A. Rohwer. 1972. Hermon Bumpus and natural selection in the house sparrow *Passer domesticus*. *Evolution* 26:20–31.

Johnston, R. F., and R. K. Selander. 1971. Evolution in the house sparrow. II. Adaptive differentiation in North American populations. *Evolution* 25:1–28.

Johnston, R. F., and R. K. Selander. 1973. Evolution in the house sparrow. III. Variation in size and sexual dimorphism in Europe and North and South America. *Amer. Nat.* 107:373–390.

Jones, D. A. 1962. Selective eating of the acyanogenic form of the plant *Lotus corniculatus* L. by various animals. *Nature* 193:1109–1110.

Jones, D. A. 1966. On the polymorphism of cyanogenesis in *Lotus corniculatus* L. I. Selection by animals. *Can. J. Genet. Cytol.* 8:556–567.

Jones, D. A. 1970. On the polymorphism of cyanogenesis in *Lotus corniculatus* L. III. Some aspects of selection. *Heredity* 25:633–641.

Jones, D. A. 1972. On the polymorphism of cyanogenesis in *Lotus corniculatus* L. IV. The Netherlands. *Genetica* 43:394–406.

Jones, J. S. 1982. Genetic differences in individual behavior associated with shell polymorphism in the snail *Cepaea nemoralis*. *Nature* 298:749–750.

Jones, J. S., and D. T. Parkin. 1977. Experimental manipulation of some snail populations subject to climatic selection. *Amer. Nat.* 111:1014–1017.

Jones, J. S. and R. F. Probert. 1980. Habitat selection maintains a deleterious allele in a heterogeneous environment. *Nature* 287:632–633.

Kahler, A. L., R. W. Allard, M. Krzakowa, C. F. Wehrhahn, and E. Nevo. 1980. Associations between isozyme phenotypes and environment in the slender wild oat (*Avena barbata*) in Israel. *Theor. Appl. Genet.* 56:31–47.

Kahler, A. L., M. T. Clegg, and R. W. Allard. 1975. Evolutionary changes in the mating system of an experimental population of barley (*Hordeum vulgare* L.). *Proc. Nat. Acad. Sci. U.S.A.* 72:943–946.

Kaplan, R. H., and W. S. Cooper. 1984. The evolution of developmental plasticity in reproductive characteristics: An application of the "adaptive coin-flipping" principle. *Amer. Nat.* 123:393–410.

Karataglis, S. S. 1980a. Gene flow in *Agrostis tenuis*. *Plant Syst. Evol.* 134:23–31.

Karataglis, S. S. 1980b. Selective adaptation to copper of populations of *Agrostis tenuis* and *Festuca rubra*. *Plant Syst. Evol.* 134:215–228.

Karlin, S., and U. Lieberman. 1974. Random temporal variation in selection intensities: Case of large population size. *Theor. Popul. Biol.* 6:355–382.

Karn, M. N., and L. S. Penrose. 1951. Birth weight and gestation time in relation to maternal age, parity, and infant survival. *Ann. Eugen.* 16:147–164.

Karson, M. J. 1982. *Multivariate Statistical Methods*. Ames: Iowa State University Press.

Kempthorne, O., and E. Pollak. 1970. Concepts of fitness in Mendelian populations. *Genetics* 64:125–145.

Kendall, W. A., and R. T. Sherwood. 1975. Palatability of leaves of tall fescue and reed canarygrass and of some of their alkaloids to meadow voles. *Agron. J.* 67:667–671.

Kettlewell, H.B.D. 1958. A survey of the frequencies of *Biston betularia* L. (Lep.) and its melanic forms in Britain. *Heredity* 12:51–72.

Kettlewell, H.B.D. 1973. *The Evolution of Melanism: The Study of a Recurring Necessity*. Oxford: Oxford University Press.

Kettlewell, H.B.D., and R. J. Berry. 1969. Gene flow in a cline. *Amathes glareosa* Esp. and its melanic f. *edda* Staud. (Lep.) in Shetland. *Heredity* 24:1–14.

Kettlewell, H.B.D., R. J. Berry, C. J. Cadbury, and G. C. Phillips. 1969. Differences in behaviour, dominance and survival within a cline. *Amathes glareosa* Esp. and its melanic f. *edda* Staud. (Lep.) in Shetland. *Heredity* 24:15–25.

Keymer, R. J., and W. M. Ellis. 1978. Experimental studies on plants of *Lotus corniculatus* L. from Anglesey, polymorphic for cyanogenesis. *Heredity* 40:189–206.

Khazaeli, A. A., and D. J. Heath. 1979. Colour polymorphism, selection, and the sex ratio in the isopod *Sphaeroma rugicauda* (Leach). *Heredity* 42:187–199.

Kiang, Y. T., and W. J. Libby. 1972. Maintenance of a lethal in a natural population of *Mimulus guttatus*. *Amer. Nat.* 106:351–367.

Kidwell, M. G. 1983a. Evolution of hybrid dysgenesis determinants in *Drosophila melanogaster*. *Proc. Nat. Acad. Sci. U.S.A.* 80:1655–1659.

Kidwell, M. G. 1983b. Intraspecific hybrid sterility. Pp. 125–154 in M. Ashburner, H. L. Carson, and J. N. Thompson, Jr., eds., *Genetics and Biology of Drosophila*. New York: Academic Press.

Kiester, A. R., R. Lande, and D. W. Schemske. 1984. Models of coevolution and speciation in plants and their pollinators. *Amer. Nat.* 124:220–243.

Kimura, M. 1954. Processes leading to quasi-fixation of genes in natural populations due to random fluctuation of selection intensities. *Genetics* 39:280–295.

Kimura, M. 1983. *The Neutral Theory of Molecular Evolution*. Cambridge, Eng.: Cambridge University Press.

Kimura, M., and J. F. Crow. 1978. Effect of overall phenotypic selection on genetic change at individual loci. *Proc. Nat. Acad. Sci. U.S.A.* 75:6168–6171.

King, M-C., and A. C. Wilson. 1975. Evolution at two levels in humans and chimpanzees. *Science* 188:107–116.

Kirkpatrick, M. 1982. Sexual selection and the evolution of female choice. *Evolution* 36:1–12.

Kirpichnikov, V. S., and I. M. Ivanova. 1978. Space, temporal, and age-dependent variation in Pacific sockeye salmon for $LDH\text{-}B^1$ and PGM-1 loci. *Genetika, Moscow* 13:1183–1193 (1977). Translated as *Soviet Gen.* 13:791–799.

Klar, G. T., C. B. Stalnaker, and T. M. Farley. 1979a. Comparative physical and physiological performance of rainbow trout, *Salmo gairdneri*, of distinct lactate dehydrogenase B^2 phenotypes. *Comp. Biochem. Physiol.* 63A:229–235.

Klar, G. T., C. B. Stalnaker, and T. M. Farley. 1979b. Comparative blood lactate response to low oxygen concentrations in rainbow trout, *Salmo gairdneri*, $LDH\text{–}B^2$ phenotypes. *Comp. Biochem. Physiol.* 63A:237–240.

Knights, R. W. 1979. Experimental evidence for selection on shell size in *Cepaea hortensis* (Mull.). *Genetica* 50:51–60.

Koehn, R. K., and F. W. Immermann. 1981. Biochemical studies of aminopeptidase polymorphism in *Mytilus edulis*. I. Dependence of enzyme activity on season, tissue, and genotype. *Biochem. Genet.* 19: 1115–1142.

Koehn, R. K., and J. B. Mitton. 1972. Population genetics of marine pelecypods. I. Ecological heterogeneity and evolutionary strategy at an enzyme locus. *Amer. Nat.* 106:47–56.

Koehn, R. K., R.I.E. Newell, and F. W. Immermann. 1980. Maintenance of an aminopeptidase allele frequency cline by natural selection. *Proc. Nat. Acad. Sci. U.S.A.* 77:5385–5389.

Koehn, R. K., and J. F. Siebenaller. 1981. Biochemical studies of aminopeptidase polymorphism in *Mytilus edulis*. II. Dependence of reaction rate on physical factors and enzyme concentration. *Biochem. Genet.* 19:1143–1162.

Koehn, R. K., F. J. Turano, and J. B. Mitton. 1973. Population genetics of marine pelecypods. II. Genetic differences in microhabitats of *Modiolus demissus*. *Evolution* 27:100–105.

Koehn, R. K., A. J. Zera, and J. G. Hall. 1983. Enzyme polymorphism and natural selection. Pp. 115–136 in M. Nei and R. K. Koehn, eds., *Evolution of Genes and Proteins*. Sunderland, Mass.: Sinauer Associates.

Komai, T. 1954. An actual instance of microevolution observed in an insect population. *Proc. Imperial Acad. Japan* 30:970–975.

Konovalov, S. M., and A. G. Shevlyakov. 1978. Natural selection in Pacific salmon (*Oncorhyncus nerka* Lalb.) by body size. *Zh. Obsch. Biol.* 39:194–206.

Krebs, J. R., and N. B. Davies, eds. 1978. *Behavioural Ecology, an Evolutionary Approach*. Sunderland, Mass.: Sinauer Associates.

Krimbas, C. B. 1976. The Lewontin and Krakauer test on quantitative characters. *Genetics* 84:395–397.

Krimbas, C. B., and S. Tsakas. 1971. The genetics of *Dacus oleae*. V. Changes of esterase polymorphism in a natural population following insecticide control—selection or drift? *Evolution* 25:454–460.

Kurtén, B. 1953. On the correlation in the mammalian dentition: Characteristics of the correlation. *Acta Zool. Fennica* No. 76:1–122.

Kurtén, B. 1955. Sex dimorphism and six trends in the cave bear, *Ursus spelaeus* Rosenmuller and Heinroth. *Acta Zool. Fennica* No. 90: 1–48.

Kurtén, B. 1957. A case of Darwinian selection in bears. *Evolution* 11:412–416.

Kurtén, B. 1958a. Life and death of a Pleistocene cave bear: A study in paleoecology. *Acta Zool. Fennica* No. 95:1–59.

Kurtén, B. 1958b. The Carnivora of the Palestine caves. *Acta Zool. Fennica* No. 107:1–74.

Lachenbruch, P. A. 1975. *Discriminant Analysis*. New York: Hafner Press.

Lachenbruch, P. A., and M. Goldstein. 1979. Discriminant analysis. *Biometrics* 35:69–85.

Lack, D. 1947a. The significance of clutch size. *Ibis* 89:302–352.

Lack, D. 1947b. The significance of clutch size in the Partridge (*Perdix perdix*). *J. Anim, Ecol.* 16:19–25.

Lacy, R. C. 1978. Dynamics of *t*-alleles in *Mus musculus* populations: Review and speculation. *Biologist* 60:41–67.

Lamotte, M. 1951. Recherches sur la structure génétique des populations naturelles de *Cepaea nemoralis* (L.). *Bull. Biol. France et Belgique* (supplement) 35:1–239.

Lande, R. 1975. The maintenance of genetic variability by mutation in a polygenic character with linked loci. *Genet. Res. Cambr.* 26:221–235.

Lande, R. 1976a. Natural selection and random genetic drift in phenotypic evolution. *Evolution* 30:314–334.

Lande, R. 1976b. The maintenance of genetic variability by mutation in a polygenic character with linked loci. *Genet. Res. Cambr.* 26:221–235.

Lande, R. 1977. Statistical tests for natural selection on quantitative traits. *Evolution* 31:442–444.

Lande, R. 1979. Quantitative genetic analysis of multivariate evolution, applied to brain-body size allometry. *Evolution* 33:402–416.

Lande, R. 1980. The genetic covariance between characters maintained by pleiotropic mutations. *Genetics* 94:203–215.

Lande, R. 1981. Models of speciation by sexual selection on polygenic traits. *Proc. Nat. Acad. Sci. U.S.A.* 79:3721–3725.

Lande, R., and S. J. Arnold. 1983. The measurement of selection on correlated characters. *Evolution* 37:1210–1226.

Leamy, L. 1978. Intensity of natural selection for odontometric traits generated by differential fertility in *Peromyscus leucopus*. *Heredity* 41: 25–34.

Leamy, L., and W. Atchley. 1984. Static and evolutionary allometry of osteometric traits in selected lines of rats. *Evolution* 398:47–54.

Lees, D. R. 1971. The distribution of melanism in the pale brindled beauty moth, *Phigalia pedaria*, in Great Britain. Pp. 152–174 in E. R. Creed, ed., *Ecological Genetics and Evolution: Essays in honour of E. B. Ford*. Oxford: Blackwell Scientific Publications.

Levin, D. A. 1969. The effect of corolla color and outline on interspecific pollen flow in *Phlox*. *Evolution* 23:444–455.

Levin, D. A., and H. W. Kerster. 1967. Natural selection for reproductive isolation in *Phlox*. *Evolution* 21:679–687.

Levin, D. A., and H. W. Kerster. 1970. Phenotype dimorphism and population fitness in *Phlox*. *Evolution* 24:128–134.

Levinton, J. S. 1982. The body size-prey size hypothesis: The adequacy of body size as a vehicle for character displacement. *Ecology* 63:869–872.

Levinton, J. S., and R. K. Koehn. 1976. Population genetics of mussels. Pp. 357–384 in B. L. Bayne, ed., *Marine Mussels: Their Ecology and Physiology*. New York: Cambridge University Press.

Levitan, M. 1973a. Studies of linkage in populations. VI. Periodic selection for X-chromosome gene arrangement combinations. *Evolution* 27:215–225.

Levitan, M. 1973b. Studies of linkage in populations. VII. Temporal variation and X-chromosome linkage disequilibriums. *Evolution* 27:476–485.

Lewontin, R. C. 1962. Interdeme selection controlling a polymorphism in the house mouse. *Amer. Nat.* 96:65–78.

Lewontin, R. C. 1970. The units of selection. *Ann. Rev. Ecol. Syst.* 1:1–16.

Lewontin, R. C. 1972. Testing the theory of natural selection. *Nature* 236:181–182.

Lewontin, R. C. 1974. *The Genetic Basis of Evolutionary Change*. New York: Columbia University Press.

Lewontin, R. C. 1979. Fitness, survival, and optimality. Pp. 3–21 in D. J. Horn, G. R. Stairs, and R. D. Mitchell, eds., *Analysis of Ecological Systems*. Columbus: Ohio State University Press.

Lewontin, R. C., and C. C. Cockerham. 1959. The goodness-of-fit test for detecting natural selection in random mating populations. *Evolution* 13:561–564.

Lewontin, R. C., and D. Cohen. 1969. On population growth in a randomly varying environment. *Proc. Nat. Acad. Sci. U.S.A.* 62: 1056–1060.

Lewontin, R. C., and L. C. Dunn. 1960. The evolutionary dynamics of a polymorphism in the house mouse. *Genetics* 45:705–722.

Lewontin, R. C., and J. Krakauer. 1973. Distribution of gene frequency as a test of the theory of the selective neutrality of polymorphisms. *Genetics* 74:175–195.

Lewontin, R. C., and J. Krakauer. 1975. Testing the heterogeneity of *F* values. *Genetics* 80:397–398.

Li, C. C. 1959. Notes on relative fitness of genotypes that forms a geometric progression. *Evolution* 13:564–567.

Lieberman, M., and D. Lieberman. 1970. The evolutionary dynamics of the desert woodrat *Neotoma lepida*. *Evolution* 24:560–570.

Lincoln, R. J., G. A. Boxshall, and P. F. Clark. 1982. *A Dictionary of Ecology, Evolution and Systematics*. Cambridge, Eng.: Cambridge University Press.

Lindsey, C. C. 1962. Experimental study of meristic variation in a population of three-spine sticklebacks, *Gasterosteus aculeatus*. *Canad. J. Zool.* 40:271–312.

Linhart, Y. B., J. B. Mitton, D. M. Bowman, and K. B. Sturgeon. 1979. Genetic aspects of fertility differentials in ponderosa pine. *Genet. Res. Cambr.* 33:237–242.

Linhart, Y. B., J. B. Mitton, K. B. Sturgeon, and M. L. Davis. 1981. Genetic variation in space and time in a population of ponderosa pine. *Heredity* 46:407–426.

Lininger, L., M. H. Gail, S. B. Green, and D. P. Byar. 1979. Comparison of four tests for equality of survival curves in the presence of stratification and censoring. *Biometrika* 66:419–428.

Lowther, P. E. 1977. Selection intensity in North American house sparrows (*Passer domesticus*). *Evolution* 31:649–656.

Luedders, V. D. 1978. Effect of planting date on natural selection in soybean populations. *Crop. Sci.* 18:943–944.

Luedders, V. D., and L. A. Duclos. 1978. Reproductive advantage associated with resistance to soybean-cyst nematode. *Crop. Sci.* 18:821–823.

Lukefahr, M. J., and J. E. Houghtaling. 1969. Resistance of cotton strains with high gossypol content to *Heliothis* spp. *J. Econ. Entomol.* 62:588–591.

Luzzatto, L., E. A. Usanga, and S. Reddy. 1969. Glucose-6-phosphate dehydrogenase deficient red cells: Resistance to infection by malarial parasites. *Science* 164:839–841.

Lynch, M. 1984. The selective value of alleles underlying polygenic traits. *Genetics* 108:1021–1033.

Lyttle, T. W. 1979. Experimental population genetics of meiotic drive systems. II. Accumulation of genetic modifiers of segregation distorter in laboratory populations. *Genetics* 91:339–357.

MacArthur, R. H. 1972. *Geographical Ecology*. New York: Harper and Row.

Macbeth, N. 1971. *Darwin Retried*. Boston: Gambit Press.

McCauley, D. E. 1979. Geographic variation in body size and its relation to the mating structure of *Tetraopes* populations. *Heredity* 42: 143–148.

McCauley, D. E., and M. J. Wade. 1978. Female choice and mating structure of a natural population of the soldier beetle, *Chauliognathus pennsylvanicus*. *Evolution* 32:771–775.

McCracken, G. F., and J. W. Bradbury. 1977. Paternity and genetic heterogeneity in the polygynous bat: *Phyllostomus hastatus*. *Science* 198:303–306.

McDonald, J. F. 1983. The molecular basis of adaptation: A critical review of relevant ideas and observations. *Ann. Rev. Ecol. Syst.* 14:77–102.

McGregor, P. K., J. R. Krebs, and C. M. Perrins, 1981. Song repertoires and lifetime reproductive success in the great tit (*Parus major*). *Amer. Nat.* 118:149–159.

McKenzie, J. A., and S. W. McKechnie. 1978. Ethanol tolerance and the ADH polymorphism in a natural population of *Drosophila melanogaster*. *Nature* 272:75–76.

McKenzie, J. A., and P. A. Parsons. 1974. Microdifferentiation in a natural population of *Drosophila melanogaster* to alcohol in the environment. *Genetics* 77:385–394.

McLain, D. K. 1982. Density-dependent sexual selection and positive phenotypic assortative mating in natural populations of the soldier beetle, *Chauliognathus pennsylvanicus*. *Evolution* 36:1227–1235.

MacLean, J. A. 1980. Ecological genetics of three-spine sticklebacks in Heinsholt Lake. *Canad. J. Zool.* 58:2026–2039.

McNeilly, T. 1968. Evolution in closely adjacent plant populations. III. *Agrostis tenuis* on a small copper mine. *Heredity* 23:99–108.

McNeilly, T. 1979. Studies on the ecological genetics of heavy metal tolerant plant populations. *Aquilo, Ser. Zool.* 20:17–25.

McNeilly, T., and A. D. Bradshaw. 1968. Evolutionary processes in populations of copper tolerant *Agrostis tenuis* Slbth. *Evolution* 22: 108–118.

McPhail, J. D. 1969. Predation and the evolution of a stickleback (*Gasterosteus*). *J. Fish. Res. Bd. Canada* 26:3183–3208.

Majerus, M., P. O'Donald, & J. Weir. 1982a. Evidence for preferential mating in *Adalia bipunctata*. *Heredity* 49:37–49.

Majerus, M., P. O'Donald, and J. Weir. 1982b. Female mating preference is genetic. *Nature* 300:521–523.

Mackie, J. L. 1978. Failures in criticism: Popper and his commentators. *Brit. J. Phil. Sci.* 29:363–387.

Malpica, J. M., and D. A. Briscoe. 1982. Multilocus nonrandom associations in *Drosophila melanogaster*. *Genet. Res. Cambr.* 39:41–61.

Manley, B.F.J. 1974. A model for certain types of selection experiments. *Biometrics* 30:281–294.

Manley, B.F.J. 1975. The measurement of the characteristics of natural selection. *Theor. Pop. Biol.* 7:288–305.

Manley, B.F.J. 1976. Some examples of double exponential fitness functions. *Heredity* 36:229–234.

Manley, B.F.J. 1977. A new index for the intensity of natural selection. *Heredity* 38:321–328.

Manton, S. M. 1956. The evolution of arthropod locomotory mechanisms, part 5: The structure, habits and evolution of the *Pselaphognatha* (*Diplopoda*). *Zool. J. Linn. Soc. Lond.* 43:153–187.

Manton, S. M. 1959. Functional morphology and taxonomic problems of Arthropoda. Pp. 23–32 in A. J. Cain, ed., *Function and Taxonomic Importance*. London: Systematics Association.

Marcus, L. F. 1964. Measurement of natural selection in natural populations. *Nature* 202:1033–1034.

Marcus, L. F. 1969. Measurement of selection using distance statistics in the prehistoric Orang-utan *Pongo pygmaeus palaeosumatrensis*. *Evolution* 23:301–307.

Marshall, H. G. 1976. Genetic changes in oat bulk populations under winter survival stress. *Crop. Sci.* 16:9–15.

Marten, G. C., R. F. Barnes, A. B. Simons, and F. J. Wooding. 1973. Alkaloids and palatability of *Phalaris arundinacea* L. *Agron. J.* 65: 199–201.

Martins, P. S., and S. K. Jain. 1980. Interpopulation variation in rose clover grown in diverse environments. *J. Heredity* 71:29–32.

Mason, L. G. 1964. Stabilizing selection for mating fitness in natural populations of *Tetraopes*. *Evolution* 18:492–497.

Mason, L. G. 1969. Mating selection in the California oak moth (Lepidoptera: Dioptidae). *Evolution* 23:55–58.

Mason, L. G. 1980. Sexual selection and the evolution of pair-bonding in soldier beetles. *Evolution* 34:174–180.

Mather, K. 1953. The genetical structure of populations. *Evolution, Symp. Soc. Exp. Biol.* 7:66–95.

May, R. M. 1973. *Stability and Complexity in Model Ecosystems*. Princeton, N.J.: Princeton University Press.

May, R. M., and R. M. Anderson. 1983. Parasite-host coevolution. Pp. 186–206 in D. J. Futuyma and M. Slatkin, eds., *Coevolution*. Sunderland, Mass.: Sinauer Associates.

Maynard Smith, J. 1978. *The Evolution of Sex*. Cambridge, Eng.: Cambridge University Press.

Maynard Smith, J., and J. Haigh. 1974. The hitch-hiking effect of a favourable gene. *Genet. Res. Cambr.* 23:23–35.

Maynard Smith, J., and R. Hoekstra. 1980. Polymorphism in a varied environment: How robust are the models? *Genet. Res. Cambr.* 35:45–57.

Mayr, E. 1942. *Systematics and the Origin of Species from the Viewpoint of a Zoologist*. New York: Columbia University Press.

Mayr, E. 1961. Cause and effect in biology. *Science* 134:1501–1506.

Mayr, E. 1962. Accident or design: The paradox of evolution. Pp. 1–14 in *The Evolution of Living Organisms*, Proc. Darwin Centenary Symp., Roy. Soc. Victoria. Melbourne: Melbourne University Press.

Mayr, E. 1963. *Animal Species and Evolution*. Cambridge, Mass: Belknap Press.

Mayr, E. 1978. Evolution. *Scientific American* 239:47–55.

Mayr, E. 1983. How to carry out the adaptationist program? *Amer. Nat.* 121:324–334.

Merrell, D. J. 1969. Natural selection in a leopard frog population. *J. Minn. Acad. Sci.* 35:86–89.

Merrell, D. J., and C. F. Rodell. 1968. Seasonal selection in the leopard frog *Rana pipiens. Evolution* 22:284–288.

Milkman, R. 1982. Toward a unified selection theory. Pp. 105–118 in R. Milkman, ed., *Perspectives on Evolution.* Sunderland, Mass.: Sinauer Associates.

Miller, R. G., Jr., G. Gong, and A. Muñoz. 1981. *Survival Analysis.* New York: Wiley.

Mitton, J. B. 1977. Shell color and pattern variation in *Mytilus edulis* and its adaptive significance. *Chesapeake Sci.* 18:387–390.

Moodie, G.E.E. 1972. Predation, natural selection, and adaptation in an unusual threespine stickleback. *Heredity* 28:155–167.

Moran, N. 1981. Intraspecific variability in herbivore performance and host quality: A field study of *Uroleucon caligatum* (Homoptera: Aphididae) and its *Solidago* hosts (Asteraceae). *Ecol. Entomol.* 6: 301–306.

Moran, P.A.P. 1964. On the nonexistence of adaptive topographies. *Ann. Hum. Genet.* 27:383–393.

Muggleton, J. 1978. Selection against the melanic morphs of *Adalia bipunctata* (two-spot ladybird): A review and some new data. *Heredity* 40:269–280.

Muggleton, J. 1979. Non-random mating in wild populations of polymorphic *Adalia bipunctata. Heredity* 42:57–65.

Muggleton, J., D. Lonsdale, and B. R. Benham. 1975. Melanism in *Adalia bipunctata* L. (col., Coccinellidae) and its relationship to atmospheric pollution. *J. Appl. Ecol.* 12:451–464.

Mukai, T., and S. Nagano. 1983. The genetic structure of natural populations of *Drosophila melanogaster.* XVI. Excess of additive genetic variance of viability. *Genetics* 105:115–134.

Murray, J. J., and B. C. Clarke. 1978. Changes in gene frequency in *Cepaea nemoralis* over 50 years. *Malacologia* 17:317–330.

Murray, N. D., J. A. Bishop, and M. R. MacNair. 1980. Melanism and predation by birds in the moths *Biston betularia* and *Phigalia pilosaria. Proc. Roy. Soc. Lond. B.* 210:276–284.

Myers, J. H. 1974. Genetic and social structure of feral house mouse populations on Grizzly Island, California. *Ecology* 55:747–759.

Nadeau, J. H., and R. Baccus. 1981. Selection components of four allozymes in natural populations of *Peromyscus maniculatus*. *Evolution* 35:11–20.

Nadeau, J. H., and R. Baccus. 1983. Gametic selection and hemoglobin polymorphisms in *Peromyscus maniculatus*: A rejoinder. *Evolution* 37:642–646.

Nei, M. 1973. The theory and estimation of genetic distance. Pp. 45–51 in N. E. Morton, ed., *Genetic Structure of Populations*. Honolulu: University of Hawaii Press.

Nei, M. 1978. The theory of genetic distance and evolution of human races. *Japan. J. Hum. Genet.* 23:341–369.

Nei, M. 1983. Genetic polymorphism and the role of mutation in evolution. Pp. 165–190 in M. Nei and R. K. Koehn, eds., *Evolution of Genes and Proteins*. Sunderland, Mass.: Sinauer Associates.

Nei, M., and A. Chakravarti. 1977. Drift variances of F_{ST} and G_{ST} statistics obtained from a finite number of isolated populations. *Theor. Pop. Biol.* 11:307–325.

Nei, M., A. Chakravarti, and Y. Tateno. 1977. Mean and variance of F_{ST} in a finite number of incompletely isolated populations. *Theor. Pop. Biol.* 11:291–306.

Nei, M., and R. K. Chesser. 1983. Estimation of fixation indices and gene diversities. *Ann. Hum. Genet.* 47:253–259.

Nei, M., P. A. Fuerst, and R. Chakraborty. 1976. Testing the neutral mutation hypothesis by distribution of single locus heterozygosity. *Nature* 262:491–493.

Nei, M., and W-H. Li 1973. Linkage disequilibrium in subdivided populations. *Genetics* 75:213–219.

Nei, M., and A. K. Roychoudhury. 1974. Sampling variances of heterozygosity and genetic distance. *Genetics* 76:379–390.

Nevo, E. 1978. Genetic variation in natural populations: Pattern and theory. *Theor. Pop. Biol.* 13:121–177.

Nevo, E. 1979. Adaptive convergence and divergence of subterranean mammals. *Ann. Rev. Ecol. Syst.* 10:269–308.

Nevo, E., H. C. Dessauer, and K-C. Chuang. 1975. Genetic variation as a test of natural selection. *Proc. Nat. Acad. Sci. U.S.A.* 72: 2145–2149.

Nevo, E., T. Shimony, and M. Libni. 1977. Thermal selection of allozyme polymorphism in barnacles. *Nature* 267:699–701.

New, J. K. 1958. A population study of *Spergula arvensis*. I. Two clines and their significance. *Ann. Bot.* 22:457–478.

New, J. K. 1959. A population study of *Spergula arvensis*. II. Genetics and breeding behaviour. *Ann. Bot.* 23:23–33.

New, J. K. 1978. Change and stability of clines in *Spergula arvensis* L. (corn spurrey) after 20 years. *Watsonia* 12:137–143.

Nunney, L. 1985. Group selection, altruism and structured-deme models. *Amer. Nat.* 126:212–230.

Oakeshott, J. G., G. K. Chambers, J. B. Gibson, and D. A. Willcocks. 1981. Latitudinal relationships of esterase-6 and phosphoglucomutase gene frequencies in *Drosophila melanogaster*. *Heredity* 47:385–396.

Oakeshott, J. G., J. B. Gibson, P. R. Anderson, W. R. Knibb, D. G. Anderson, and G. K. Chambers. 1982. Alcohol dehydrogenase and glycerol-3-phosphate dehydrogenase clines in *Drosophila melanogaster* on different continents. *Evolution* 36:86–96.

O'Donald, P. 1968. Measuring the intensity of natural selection. *Nature* 220:197–198.

O'Donald, P. 1970. Change of fitness by selection for a quantitative character. *Theor. Popul. Biol.* 1:219–232.

O'Donald, P. 1971. Natural selection for quantitative characters. *Heredity* 27:137–153.

O'Donald, P. 1972. Natural selection of reproductive rates and breeding times and its effect on sexual selection. *Amer. Nat.* 106: 368–379.

O'Donald, P. 1973. A further analysis of Bumpus's data: The intensity of natural selection. *Evolution* 27:398–404.

O'Donald, P. 1983. Sexual selection and fertility. *Theor. Pop. Biol.* 23:64–84.

O'Donald, P., and J.W.F. Davis. 1975. Demography and selection in a population of Arctic Skuas. *Heredity* 35:75–83.

O'Donald, P., and P. E. Davis. 1959. The genetics of the color phases of the Arctic Skua. *Heredity* 13:481–486.

O'Donald, P., M. Derrick, M. Majerus, and J. Weir. 1984. Population genetics theory of the assortative mating, sexual selection and natural selection of the two-spot ladybird *Adalia bipunctata*. *Heredity* 52:43–61.

O'Donald, P., N. S. Wedd, and J.W.F. Davis. 1974. Mating preferences and sexual selection in the Arctic Skua. *Heredity* 33:1–16.

Olson, S. L., and A. Feduccia. 1980. Relationships and evolution of flamingos (Aves: Phoenicopteridae). *Smithsonian Contr. Zool.* 316:1–73.

Oslon, S. L., and Y. Hasegawa. 1979. Fossil counterparts of giant penguins from the North Pacific. *Science* 206:688–689.

Oster, G. F., and E. O. Wilson. 1978. *Caste and Ecology in the Social Insects.* Princeton, N.J.: Princeton University Press.

Ostergaard, H., and F. B. Christiansen. 1981. Selection component analysis of natural populations using population samples including mother-offspring combinations. II. *Theor. Pop. Biol.* 19:378–419.

Otte, D., and K. Williams. 1972. Environmentally induced color dimorphisms in grasshoppers, *Syrbula admirabilis Dichromorpha viridis* and *Chortophaga viridifasciata. Ann. Entomol. Soc. Amer.* 65:1154–1161.

Oxford, G. S. 1977. Multiple sources of esterase enzymes in the crop juice of *Cepaea* (Mollusca: Helicidae). *J. Comp. Physiol.* 122:375–383.

Oxford, G. S. 1978. The nature and distribution of food-induced esterases in helicid snails. *Malacologia* 17:331–339.

Oxford, G. S., and T. Andrews. 1977. Variation in characters affecting fitness between radiate and non-radiate morphs in natural populations of groundsel (*Senecio vulgaris* L.). *Heredity* 38:367–371.

Packard, A. 1972. Cephalopods and fish: The limits of convergence. *Biol. Rev.* 47:241–307.

Palenzona, D. L., C. Fini, and R. E. Scossiroli. 1971. Comparative study of natural selection in *Cardium edulae* L. and *Drosophila melanogaster* Meig. *Monit. Zool. Ital.*, n.s., 5:165–172.

Paquin, C. E., and J. Adams. 1983. Relative fitness can decrease in evolving asexual populations of *S. cerevisiae. Nature* 306:368–371.

Parker, G. A. 1983. Mate quality and mating decisions. Pp. 141–166 in P. Bateson, ed., *Mate Choice.* Cambridge, Eng.: Cambridge University Press.

Partridge, G. G. 1979. Relative fitness of genotypes in a population of *Rattus norvegicus* polymorphic for Warfarin resistance. *Heredity* 43:239–246.

Patton, J. L., R. J. Baker, and H. H. Genoways. 1980. Apparent chromosomal heterosis in a fossorial mammal. *Amer. Nat.* 116:143–146.

Pearson, B. 1981. Matter of principle. *Nature* 293:6.

Pearson, K. 1903. Mathematical contributions to the theory of evolution. XI. On the influence of natural selection on the variability and correlation of organs. *Phil. Trans. Roy. Soc. Lond. A.* 200:1–66.

Perrins, C. M. 1965. Population fluctuations and clutch size in the great tit, *Parus major* L. *J. Anim. Ecol.* 34:601–647.

Perzigian, A. J. 1975. Natural selection on the dentition of an Arikara population. *Amer. J. Phys. Anthropol.* 42:63–70.

Peters, R. H. 1976. Tautology in evolution and ecology. *Amer. Nat.* 110:1–12.

Piazza, A., P. Menozzi, and L. L. Cavalli-Sforza. 1981. Synthetic gene frequency maps of man and selective effects of climate. *Proc. Nat. Acad. Sci. U.S.A.* 78:2638–2642.

Place, A. R., and D. A. Powers. 1979. Genetic variation and relative catalytic efficiencies: Lactate dehydrogenase B allozymes of *Fundulus heteroclitus*. *Proc. Nat. Acad. Sci. U.S.A.* 76:2354–2358.

Platnick, N. 1977. Evolution and the diversity of life: Selected essays. *Syst. Zool.* 26:224–228.

Popescu, C. 1979. Natural selection in the industrial melanic psocid *Mesopsocus unipunctatus* (Mull.) (Insecta: Psocoptera) in northern England. *Heredity* 42:133–142.

Popper, K. 1963. *The Poverty of Historicism*. London: Routledge & Kegan Paul.

Popper, K. 1972. *Objective Knowledge*. Oxford: Oxford University Press.

Popper, K. 1974. *The Philosophy of Karl Popper*, ed. P. A. Schilpp, vol. 1, pp. 133–143. La Salle, Ill.: Open Court Press.

Popper, K. 1978. Natural selection and the emergence of mind. *Dialectica* 32:339–355.

Potts, D. C. 1978. Differentiation in coral populations. *Atoll Res. Bull.* 220:55–74.

Potts, D. C. 1984. Natural selection in experimental populations of reef-building corals (*Scleractinia*). *Evolution* 38:1059–1078.

Powers, D. A., and A. R. Place. 1978. Biochemical genetics of *Fundulus heteroclitus* (L.). I. Temporal and spatial variation in gene frequencies of LDH-B, MDH-A, GPI-B, and PGM-A. *Biochem. Genet.* 16:593–607.

Powers, D. A., G. S. Greaney, and A. R. Place. 1979. Physiological correlation between lactate dehydrogenase genotype and haemoglobin function in killifish. *Nature* 277:240–241.

Prakash, S. 1977. Gene polymorphism in natural populations of *Drosophila persimilis*. *Genetics* 85:513–520.

Prakash, S., and R. C. Lewontin. 1971. A molecular approach to the study of genic heterozygosity in natural populations. V. Further direct evidence of coadaptation in inversions of *Drosophila*. *Genetics* 69:405–408.

Price, T. D. 1984. Sexual selection on body size, territory, and plumage variables in a population of Darwin's finches. *Evolution* 38: 327–341.

Price, T. D., and P. R. Grant. 1984. Life history traits and natural selection for small body size in a population of Darwin's finches. *Evolution* 38:483–494.

Price, T. D., P. R. Grant, H. L. Gibbs, and P. T. Boag. 1984. Recurrent patterns of natural selection in a population of Darwin's finches. *Nature* 309:787–791.

Prout, T. 1965. The estimation of fitnesses from genotype frequencies. *Evolution* 19:546–551.

Prout, T. 1969. The estimation of fitnesses from population data. *Genetics* 63:949–967.

Prout, T. 1971a. The relation between fitness components and population prediction in *Drosophila*. I. The estimation of fitness components. *Genetics* 68:127–149.

Prout, T. 1971b. The relation between fitness components and population prediction in *Drosophila*. II. Population prediction. *Genetics* 68:151–167.

Quisenberry, J. E., B. Roark, J. D. Bilbro, and L. L. Ray. 1978. Natural selection in a bulked hybrid population of upland cotton. *Crop. Sci.* 18:799–801.

Raff, R. A., and T. C. Kaufman. 1983. *Embryos, Genes, and Evolution: The Developmental-Genetic Basis of Evolutionary Change*. New York: Macmillan.

Rausher, M. D. 1984. The evolution of habitat preference in subdivided populations. *Evolution* 38:596–608.

Redfield, J. A. 1973a. The use of incomplete family data in the analysis of genetics and selection at the *Ng* locus in blue grouse (*Dendragapus obscurus*). *Heredity* 31:35–42.

Redfield, J. A. 1973b. Demography and genetics in colonizing populations of blue grouse (*Dendragapus obscurus*). *Evolution* 27:576–592.

Redfield, J. A. 1974. Genetics and selection at the *Ng* locus in the blue grouse (*Dendragapus obscurus*). *Heredity* 33:69–78.

Reed, E. S. 1981. The lawfulness of natural selection. *Amer. Nat.* 118:61–71.

Reed, T. E. 1968. Research on blood groups and selection from the child health and development studies, Oakland, California. II. Gravidae's characteristics. *Amer. J. Hum. Genet.* 20:119–128.

Rendel, J. M. 1943. Variation in the weights of hatched and unhatched duck's eggs. *Biometrika* 33:48–56.

Rensch, B. 1959. *Evolution above the Species Level.* New York: Columbia University Press.

Reynolds, J., B. S. Weir, and C. C. Cockerham. 1983. Estimation of the coancestry coefficient: Basis for a short-term genetic distance. *Genetics* 105:767–779.

Reznick, D. A. 1982. The impact of predation on life history evolution in Trinidadian guppies: Genetic basis of observed life history patterns. *Evolution* 36:1236–1250.

Reznick, D., and J. A. Endler. 1982. The impact of predation on life history evolution in Trinidadian guppies (*Poecilia reticulata*). *Evolution* 36:160–177.

Richardson, A.M.M. 1974. Differential climatic selection in natural populations of the land snail *Cepaea nemoralis*. *Nature* 247:572–573.

Riddell, B. E., and W. C. Leggett. 1981. Evidence of an adaptive basis for geographic variation in body morphology and time of downstream migration of juvenile Atlantic salmon (*Salmo salar*). *Can. J. Fish. Aquat. Sci.* 38:308–320.

Riddell, B. E., W. C. Leggett, and R. L. Saunders. 1981. Evidence for adaptive polygenic variation between two populations of Atlantic salmon (*Salmo salar*) native to tributaries of the S. W. Miramache River, N. B. *Can. J. Fish. Aquat. Sci.* 38:321–333.

Robertson, A. 1975. Remarks on the Lewontin-Krakauer test. *Genetics* 80:396.

Robertson, A., ed. 1980. *Selection Experiments in Laboratory and Domestic Animals*. Proceedings of a symposium held at Harrogate, U.K., on 21–22 July 1979. Commonwealth Agricultural Bureau, Slough, U.K.

Robertson, A., and W. G. Hill. 1984. Deviations from Hardy-Weinberg proportions: Sampling variances and use in estimation of inbreeding coefficients. *Genetics* 107:703–718.

Robson, G. C., and O. W. Richards. 1936. *The Variation of Animals in Nature*. London: Longmans, Green & Co.

Robson, M. 1981. Non-random survival. *Nature* 293:594.

Rosen, D. E. 1978. Review of *Introduction to Natural Selection* by Johnson. *Syst. Zool.* 27:370–373.

Rosen, D. E., and D. G. Buth. 1980. Empirical evolutionary research versus neo-Darwinian speculation. *Syst. Zool.* 29:300–308.

Roughgarden, J. 1971. Density-dependent natural selection. *Ecology* 52:453–468.

Ruse, M. 1977. Karl Popper's philosophy of biology. *Phil. Sci.* 44:638–661.

Sacks, J. M. 1967. A stable equilibrium with minimum average fitness. *Genetics* 56:705–708.

Sactor, B. 1975. Biochemistry of insect flight. Part 1. Utilization of fuels by muscle. Pp. 1–88 in D. Candy and B. Kilby, eds., *Insect Biochemistry and Function*. London: Chapman & Hall.

Salmon, W. C. 1971. *Statistical Explanation and Statistical Relevance*. Pittsburgh, Pa.: University of Pittsburgh Press.

Samallow, P. B. 1980. Selection, mortality, and reproduction in a natural population of *Bufo boreas*. *Evolution* 34:18–39.

Sambol, M., and R. M. Finks. 1977. Natural selection in a Cretaceous oyster. *Paleobiology* 3:1–16.

Schaal, B. A., and D. A. Levin. 1976. The demographic genetics of *Liatris cylindracea* Michx. (Compositae). *Amer Nat.* 110:191–206.

Schaap, T. 1980. The application of the Hardy-Weinberg principle to the study of populations. *Ann. Hum. Genet.* 44:211–215.

Scheiring, J. F. 1977. Stabilizing selection for size as related to mating fitness in *Tetraopes*. *Evolution* 31:447–449.

Schluter, D. 1984. Morphological and phylogenetic relations among the Darwin's finches. *Evolution* 38:921–930.

Schluter, D., T. D. Price, and P. R. Grant. 1985. Ecological character displacement in Darwin's finches. *Science* 227:1056–1058.

Schoener, T. W. 1979. Inferring the properties of predation and other injury-producing agents from injury frequencies. *Ecology* 60:1110–1115.

Schoener, T. W. 1983. Field experiments on interspecific competition. *Amer. Nat.* 122:240–285.

Schoener, T. W., and A. Schoener. 1980. Ecological and demographic correlates of injury rates in some Bahamian *Anolis* lizards. *Copeia* 1980:839–850.

Schopf, T. M., M. D. Ohman, and R. Bleiweiss. 1975. Significant age-dependent and locality-dependent changes occur in gene frequencies in the ribbed mussel *Modiolus demissus* from a single salt marsh. *Biol. Bull. Woods Hole* 149:446.

Service, P. M., and M. Rose. 1985. Genetic correlations among life-history characters: The effect of novel environments. *Evolution* 39: 943–945.

Shami, S. A., and A. M. Tahir. 1979. Operation of natural selection on human height. *Pakistan J. Zool.* 11:75–83.

Sheppard, P. M. 1951a. Fluctuations in the selective values of certain phenotypes in the polymorphic land snail *Cepaea nemoralis*. *Heredity* 5:125–134.

Sheppard, P. M. 1951b. A quantitative study of two populations of the moth *Panaxia dominula* (L.). *Heredity* 5:349–378.

Sheppard, P. M., and L. M. Cook. 1962. The manifold effects of the *Medionigra* gene of the moth *Panaxia dominula* and the maintenance of a polymorphism. *Heredity* 17:415–426.

Sherry, D. F., and B. G. Galef, Jr. 1984. Cultural transmission without imitation: Milk bottle opening by birds. *Anim. Behav.* 32:937–938.

Siegel, S. 1956. *Nonparametric Statistics for the Behavioral Sciences*. New York: McGraw-Hill.

Simons, A. B., and G. C. Marten. 1971. Relationship of indole alkaloids to palatability of *Phalaris arundinacea* L. *Agron. J.* 63:915–919.

Simpson, G. G. 1944. *Tempo and Mode in Evolution*. New York: Columbia University Press.

Simpson, G. G. 1949. *The Meaning of Evolution*. New Haven, Conn.: Yale University Press.

Singh, L. N., and L.P.V. Johnson. 1969. Natural selection in a composite cross of barley. *Can. J. Genet. Cytol.* 11:34–42.

Slatkin, M. 1980. Ecological character displacement. *Ecology* 61:163–177.

Slatkin, M. 1983. Genetic background. Pp. 14–32 in D. J. Futuyma and M. Slatkin, eds., *Coevolution*. Sunderland, Mass.: Sinauer Associates.

Snaydon, R. W. 1970. Rapid population differentiation in a mosaic environment. I. The response of *Anthoxanthum odoratum* to soils. *Evolution* 24:257–269.

Snaydon, R. W., and M. S. Davies. 1972. Rapid population differentiation in a mosaic environment. II. Morphological variation in *Anthoxanthum odoratum*. *Evolution* 26:390–405.

Snedecor, G. W., and W. G. Cochran. 1967. *Statistical Methods*. 6th ed. Ames: Iowa State University Press.

Snyder, L.R.G. 1983. Selection components affecting hemoglobins in *Peromyscus maniculatus*: A re-evaluation. *Evolution* 37:639–642.

Sober, E. 1981. Holism, individualism and the units of selection. Pp. 93–121 in P. D. Asquith and R. N. Giere, eds., *P.S.A. 1980*, vol. 2. East Lansing, Mich.: Philosophy of Science Association.

Sober, E. 1984. *The Nature of Selection: A Philosphical Inquiry*. Cambridge, Mass.: Bradford/M.I.T. Press.

Sokal, R. R., and F. J. Rohlf. 1981. *Biometry*. 2d ed. San Francisco: W. H. Freeman.

Southwood, T.R.E. 1978. *Ecological Methods*. 2d ed. London: Chapman & Hall.

Southwood, T.R.E. 1981. Bionomic strategies and population parameters. Pp. 30–52 in R. M. May, ed., *Theoretical Ecology, Principles and Applications*. 2d ed. Sunderland, Mass.: Sinauer Associates.

Spuhler, J. M. 1976. The maximum opportunity for natural selection in some human populations. Pp. 185–226 in E.B.W. Zubrow, ed., *Demographic Anthropology*. Albuquerque: University of New Mexico Press.

Stace, C. A. 1977. The origin of radiate *Senecio vulgaris* L. *Heredity* 39:383–388.

Stanley, S. M. 1979. *Macroevolution, Pattern and Process*. San Francisco: W. H. Freeman.

Stanley, S. M. 1982. Macroevolution and the fossil record. *Evolution* 36:460–473.

Stebbins, G. L. 1977. In defense of evolution: Tautology or theory? *Amer. Nat.* 111:386–390.

Stephenson, D. G. 1981. Fit for what? *Nature* 292:8, 95.

Stern, J. T., Jr. 1970. The meaning of "adaptation" and its relation to the phenomenon of natural selection. *Evol. Biol.* 4:39–66.

Steward, R. C. 1977. Melanism and selective predation in three species of moths. J. Anim. Ecol. 46:483–496.

Struhsaker, J. W. 1968. Selection mechanisms associated with intraspecific shell variation in *Littorina picta* (Prosobranchia: Mesogastropoda). *Evolution* 22:459–480.

Sturgeon, K. B. 1979. Monoterpene variation in ponderosa pine xylem resin related to western pine beetle predation. *Evolution* 33:803–814.

Sved, J. A. 1968. The stability of linked systems of loci with small population size. *Genetics* 59:543–563.

Tallamy, D. W., and R. F. Denno. 1981. Alternative life history patterns in risky environments: An example from lacebugs. Pp. 129–147 in R. F. Denno and H. Dingle, eds., *Insect Life History Patterns: Habitat and Geographic Variation.* New York: Springer-Verlag.

Tave, D. 1984. Genetics of dorsal fin ray numbers in the guppy, *Poecilia reticulata. Copeia* 1974:140–144.

Templeton, A. R. 1982. Adaptation and the integration of evolutionary forces. Pp. 15–31 in R. Milkman, ed., *Perspectives on Evolution.* Sunderland, Mass.: Sinauer Associates.

Thoday, J. M. 1953. Components of fitness. *Evolution, Symp. Soc. Exp. Biol.* 7:96–113.

Thoday, J. M. 1958. Natural selection and biological progress. Pp. 313–333 in S. A. Barnett, ed., *A Century of Darwin.* Cambridge, Mass.: Harvard University Press.

Thompson, G. 1977. The effect of a selected locus on linked neutral loci. *Genetics* 85:753–788.

Thornhill, R. 1983. Cryptic female choice and its implications in the scorpionfly *Harpobittacus nigriceps. Amer. Nat.* 122:765–788.

Thornhill, R. 1984. Alternative female choice tactics in the scorpionfly *Hylobittacus apicalis* (Mecoptera) and their implications. *Amer. Zool.* 24:367–383.

Tilling, S. M. 1983. An experimental investigation of the behaviour and mortality of artificial and natural morphs of *Cepaea nemoralis* (L.). *Biol. J. Linn. Soc. Lond.* 19:35–50.

Tinkle, D. W., and R. K. Selander. 1973. Age dependent allozymic variation in a natural population of lizards. *Biochem. Genet.* 8:231–237.

Tonzetich, J., and C. L. Ward. 1973. Adaptive chromosomal polymorphism in *Drosophila melanica*. *Evolution* 27:486–494.

Tsakas, S., and C. B. Krimbas. 1976. Testing the heterogeneity of *F* values: A suggestion and a correction. *Genetics* 83:399–401.

Tuomi, J. 1981. Structure and dynamics of Darwinian evolutionary theory. *Syst. Zool.* 30:22–31.

Turelli, M., and L. R. Ginsburg. 1983. Should individual fitness increase with heterozygosity? *Genetics* 104:191–209.

Tzoneva, M., A. G. Bulanov, M. Mavrudieva, S. Lalchev, D. Toncheva, and D. Tanev. 1980. Frequency of glucose-6-phosphate dehydrogenase in relation to altitude: A malaria hypothesis. *Bull. World Health Org.* 58:659–662.

Ulizzi, L., M. F. Gravina, and L. Terrenato. 1981. Natural selection associated with birth weight. II. Stabilizing and directional components. *Ann. Hum. Genet.* 45:207–212.

Urbach, P. 1978. Is any of Popper's argument against historicism valid? *Brit. J. Phil. Sci.* 29:117–130.

Valentine, J. W. 1976. Genetic strategies of adaptation. Pp. 78–94 in F. J. Ayala, ed., *Molecular Evolution*. Sunderland, Mass.: Sinauer Associates.

Van de Geer, J. P. 1971. *Introduction to Multivariate Analysis for the Social Sciences*. San Francisco: W. H. Freeman & Co.

Van Valen, L. 1963a. Selection in natural populations: *Merychippus primus*, a fossil horse. *Nature* 197:1181–1183.

Van Valen. L. 1963b. Selection in natural populations: Human fingerprints. *Nature* 200:1237–1238.

Van Valen, L. 1965a. Selection in natural populations. III. Measurement and estimation. *Evolution* 19:514–528.

Van Valen, L. 1965b. Selection in natural populations. IV. British housemice (*Mus musculus*). *Genetics* 36:119–134.

Van Valen, L. 1967. Selection in natural populations. VI. Variation genetics and more graphs for estimation. *Evolution* 21:402–406.

Van Valen, L. 1973. A new evolutionary law. *Evol. Theory* 1:1–30.

Van Valen, L. 1978. The statistics of variation. *Evol. Theory* 4:33–43.

Van Valen, L., and G. W. Mellin. 1967. Selection in natural populations. VII. New York babies (fetal life study). *Ann. Hum. Genet.* 31: 109–127.

Van Valen, L., and R. Weiss. 1966. Selection in natural populations. V. Indian rats (*Rattus rattus*). *Genet. Res. Cambr.* 8:261–267.

Varvio-Aho, S.-L., O. Jarvinen, K. Vepsalainen, and P. Pamilo. 1979. Seasonal changes of the enzyme gene pool in water-striders (Gerris). *Hereditas* 90:11–20.

Wade, M. J. 1978. A critical review of the models of group selection. *Quart. Rev. Biol.* 53:101–113.

Wade, M. J., and S. J. Arnold. 1980. The intensity of sexual selection in relation to male sexual behaviour, female choice, and sperm precedence. *Anim. Behav.* 28:446–461.

Wade, M. J., and D. E. McCauley. 1984. Group selection: The interaction of local deme size and migration in the differentiation of small populations. *Evolution* 38:1047–1058.

Waddington, C. H. 1960. Evolutionary adaptations. Pp. 381–402 in S. Tax, ed., *Evolution after Darwin*, vol. 1., *The Evolution of Life*. Chicago: University of Chicago Press.

Wall, S., M. A. Carter, and B. C. Clarke. 1980. Temporal changes of gene frequencies in *Cepaea hortensis*. *Biol. J. Linn. Soc. Lond.* 14: 303–317.

Wallace, B. 1958. The comparison of observed and calculated zygotic distributions. *Evolution* 12:113–115.

Wallace, B. 1981. *Basic Population Genetics*. New York: Colombia University Press.

Walsh, J. B. 1984. Hard lessons for soft selection. *Amer. Nat.* 124: 518–526.

Ward, R. H., and C. F. Sing. 1970. A consideration of the power of the χ^2 test to detect inbreeding effects in natural populations. *Amer. Nat.* 104:355–366.

Ware, D. M. 1975. Growth, metabolism, and optimum swimming speed of a pelagic fish. *J. Fish. Res. Bd. Canada* 32:33–41.

Ware, D. M. 1978. Bioenergetics of pelagic fish: Theoretical change in swimming speed and ration with body size. *J. Fish. Res. Bd. Canada* 35:220–228.

Ware, D. M. 1982. Power and evolutionary fitness of teleosts. *Canadian J. Fish. Aquatic Sci.* 39:3–13.

Waser, N. M., and M. V. Price. 1981. Pollinator choice and stabilizing selection for flower color in *Delphinium nelsonii*. *Evolution* 35:376–390.

Waser, N. M., and M. V. Price. 1983. Pollinator behaviour and natural selection for flower color in *Delphinium nelsonii*. *Nature* 302:422–424.

Wassermann, G. D. 1978. Testability of the role of natural selection within theories of population genetics and evolution. *Brit. J. Phil. Sci.* 29:223–242.

Wassermann, G. D. 1981a. On the nature of the theory of evolution. *Phil. Sci.* 48:416–437.

Wassermann, G. D. 1981b. Thoughts on Popper. *Nature* 293:6.

Watson, D. L., and A.W.A. Brown, eds. 1977. *Pesticide Management and Insecticide Resistance*. New York: Academic Press.

Watt, W. B. 1968. Adaptive significance of pigment polymorphisms in *Colias* butterflies. I. Variation of melanin pigment in relation to thermoregulation. *Evolution* 22:437–458.

Watt, W. B. 1977. Adaptation at specific loci. I. Natural selection on phosphoglucose isomerase of *Colias* butterflies: Biochemical and population aspects. *Genetics* 87:177–194.

Watt, W. B. 1983. Adaptation at specific loci. II. Demographic and biochemical elements in the maintenance of the *Colias* PGI polymorphism. *Genetics* 103:691–724.

Watt, W. B., R. C. Cassin, and M. S. Swan. 1983. Adaptation at specific loci. III. Field behavior and survivorship differences among *Colias* PGI genotypes are predictable from *in vitro* biochemistry. *Genetics* 103:725–739.

Watterson, G. A. 1978. The homozygosity test of neutrality. *Genetics* 88:405–417.

Webb, P. W. 1975. Hydrodynamics and energetics of fish propulsion. Canadian Department of the Environment, Fisheries and Marine Service, *Bull.* 190, *Fish. Res. Bd. Canada*, 158 pp.

Weir, B. S. 1979. Inferences about linkage disequilibrium. *Biometrics* 35:235–254.

Weir, B. S., R. W. Allard, and A. L. Kahler. 1972. Analysis of complex allozyme polymorphisms in a barley population. *Genetics* 72: 505–523.

Weir, B. S., R. W. Allard, and A. L. Kahler. 1974. Further analysis of complex allozyme polymorphisms in a barley population. *Genetics* 78:911–919.

Weir, B. S., and C. C. Cockerham. 1978. Testing hypotheses about linkage disequilibrium with multiple alleles. *Genetics* 88:633–642.

Weir, B. S., and C. C. Cockerham. 1979. Estimation of linkage disequilibrium in randomly mating populations. *Heredity* 42:105–111.

Weldon, W.F.R. 1899. Presidential address. *Brit. Assoc. Adv. Sci.*, Report of meeting, Bristol, 1898. *Transactions*, sec. D, pp. 887–902.

Weldon, W.F.R. 1901. A first study of natural selection in *Clausilia laminata* (Montagu). *Biometrika* 1:109–124.

Weldon, W.F.R. 1903. Note on a race of *Clausilia itala* (von Martens). *Biometrika* 3:299–307.

Whelan, R. J. 1982. Response of slugs to unacceptable food items. *J. Appl. Ecol.* 19:79–87.

Whitman, R. J. 1973. Herbivore feeding and cyanogenesis in *Trifolium repens* L. *Heredity* 30:241–245.

Whitten, M. J., J. M. Dearn, and J. A. McKenzie. 1980. Field studies on insecticide resistance in the Australian sheep blowfly, *Lucilia cuprina*. *Aust. J. Biol. Sci.* 33:725–735.

Wicklund, C. 1975. Pupal colour polymorphism in *Papilio machaon* L. and the survival in the field of cryptic vs. non-cryptic pupae. *Trans. Roy. Entomol. Soc. Lond.* 127:73–84.

Wiley, E. O. 1981. *Phylogenetics, the Theory and Practice of Phylogenetic Systematics*. New York: Wiley.

Willey, A. 1911. *Convergence in Evolution*. London: John Murray.

Williams, G. C. 1966. *Adaptation and Natural Selection*. Princeton, N.J.: Princeton University Press.

Williams, M., R. F. Barnes, and J. M. Cassady. 1971. Characterization of alkaloids in palatable and unpalatable clones of *Phalaris arundincacea* L. *Crop. Sci.* 11:213–217.

Williams, M. B. 1970. Deducing the consequences of evolution: A mathematical model. *J. Theor. Biol.* 29:343–385.

Williams, M. B. 1973. Falsifiable predictions of evolutionary theory. *Philosophy of Science* 40:518–537.

Wilson, D. S. 1983. The group selection controversy: History and current status. *Ann. Rev. Ecol. Syst.* 14:159–187.

Wimbush, M., L. Nemeth, and B. Birdsall. 1982. Current-induced sediment movement in the deep Florida straits: Observations. Pp. 77–94 in K. A. Fanning and F. T. Manheim, eds., *The Dynamic Environment of the Ocean Floor*. Lexington, Mass.: Lexington Books.

Wimsatt, W. C. 1981. The units of selection and the structure of the multi-level genome. Pp. 122–183 in P. D. Asquith and R. N. Giere, eds., *P.S.A. 1980*, vol. 2. East Lansing, Mich.: Philosophy of Science Association.

Wolda, H. 1963. Natural populations of the polymorphic snail *Cepaea nemoralis* (L.) *Arch. Neerl. Zool.* 15:381–471.

Wolda, H. 1967. The effect of temperature on reproduction in some morphs of the landsnail *Cepaea nemoralis* (L.). *Evolution* 21:117–129.

Workman, P. L. 1969. The analysis of simple genetic polymorphisms. *Hum. Biol.* 41:97–114.

Wright, S. 1931. Evolution in Mendelian populations. *Genetics* 16:97–159.

Wright, S. 1942. Statistical genetics and evolution. *Bull. Amer. Math. Soc.* 48:223–246.

Wright, S. 1948. On the role of directed and random changes in gene frequency in the genetics of populations. *Evolution* 2:279–294.

Wright, S. 1965. Factor interaction and linkage in evolution. *Proc. Roy. Soc. Lond. B.* 162:80–104.

Wright, S. 1967. "Surfaces" of selective value. *Proc. Nat. Acad. Sci. U.S.A.* 58:165–172.

Wright, S. 1978. *Evolution and the Genetics of Populations*, vol. 4, *Variability Within and Among Natural Populations*. Chicago: University of Chicago Press.

Wright, S. 1982. Character change, speciation, and the higher taxa. *Evolution* 36:427–443.

Zera, A. J., and R. K. Koehn, and J. G. Hall. 1985. Allozymes and biochemical adaptation. Pp. 633–674 in G. A. Kerkut and L. I. Gilbert, eds., *Comprehensive Insect Physiology*. New York: Pergamon Press.

Zink, R. M. Evolutionary and systematic significance of temporal variation in the fox sparrow. *Syst. Zool.* 32:223–238.

Species Index

(also an English name glossary)

Subject Index

(does not include section headings or chapter summaries)